Signal Traffic

THE GEOPOLITICS OF INFORMATION

Edited by Dan Schiller, Pradip Thomas,
and Yuezhi Zhao

A list of books in the series
appears at the end of this book.

Signal Traffic

Critical Studies of
Media Infrastructures

Edited by
LISA PARKS AND
NICOLE STAROSIELSKI

UNIVERSITY OF ILLINOIS PRESS
Urbana, Chicago, and Springfield

1 2 3 4 5 C P 5 4 3 2 1
This book is printed on acid-free paper.
Library of Congress Cataloging-in-Publication Data
Signal traffic: critical studies of media infrastructures / edited by
Lisa Parks and Nicole Starosielski.
pages   cm. — (The geopolitics of information)
Includes bibliographical references and index.
ISBN 978-0-252-03936-2 (hardback : acid-free paper)
ISBN 978-0-252-08087-6 (paper : acid-free paper)
ISBN 978-0-252-08087-6 (e-book)
1. Telecommunication systems—Social aspects. 2. Digital
media—Social aspects. 3. Mass media—Social aspects.
4. Information superhighway. 5. Computer networks—
Social aspects. 6. Information networks—Social aspects.
7. Telecommunication—Traffic. 8. Signal processing.
I. Parks, Lisa. II. Starosielski, Nicole, 1984–
TK5102.5.S5434   2015
303.48'33—dc23   2014049422

Contents

Acknowledgments

Signal Traffic: Critical Studies of Media Infrastructures emerged through a series of dialogues between media and communication scholars, artists, historians, digital humanities specialists, geographers, and computer scientists over the past several years. Generous grants from the University of California Humanities Research Institute (UCHRI) and the University of California Institute for Research in the Arts (UCIRA) enabled us to deepen and expand these discussions by convening two Signal Traffic workshops at UC Santa Barbara in 2010–11. We thank UCHRI director David Theo Goldberg and UCIRA director Dick Hebdige for supporting this project and for their visionary leadership of interdisciplinary humanities and arts research in the UC system. We are also grateful to the Signal Traffic workshops' participants for their stimulating presentations and contributions. These participants include: Lisa Jevbratt, Lan Xuan Le, Shannon Mattern, Tara McPherson, Toby Miller, Rahul Mukherjee, Joshua Neves, Katy Pearce, Marko Peljhan, Rita Raley, Christian Sandvig, Jonathan Sterne, Athena Tan, Lindsay Thomas, Kazys Varnelis, and Bijan Yashar. During these workshops Jon Jablonski, head of the Map and Imagery Laboratory at UC Santa Barbara, gave us a tour and delivered an eye-opening presentation on the history of cartography and infrastructure, and Kevin Schmidt in the Office of Information Technology provided a detailed overview of UC Santa Barbara's Internet history and backbone design as well as a guided tour of the hub. We thank the

Center for Information Technology and Society at UC Santa Barbara for hosting the Signal Traffic workshops and our wonderful graduate assistants, Sarah Harris and Anastasia Hill, for participating in the workshops, documenting our discussions, and ensuring everything went smoothly.

The ideas shaping this book evolved further through a panel at the 2012 Society for Cinema and Media Studies conference, in which Jonathan Sterne and Shannon Mattern participated, and during a workshop on media, materiality, and infrastructure at New York University, co-convened with Arjun Appadurai in 2014. We thank the Film and Media Studies Department at UC Santa Barbara and the Media, Culture, and Communication Department at New York University for their support in hosting and coordinating these workshops. We also thank our faculty friends, colleagues, and graduate students in these departments for the rich intellectual environments they provide daily.

As our ideas for the *Signal Traffic* collection began to cohere, we were fortunate to meet editor Daniel Nasset, who has been supportive, thoughtful, and responsive throughout the editorial process. We are grateful for his commitment to this project and the timely manner in which he helped it come to fruition. We also thank Jennifer Clark for her sharp editorial assistance. We thank this book's contributors for their field-expanding chapters and collegiality. Our deepest gratitude goes to Sarah Harris, who worked tirelessly as an editorial assistant during various stages of manuscript preparation. Finally, the project benefited from the insightful reviews of two anonymous readers, whose comments were enormously helpful as we moved the project toward the homestretch.

Beyond those mentioned above, we would like to thank Lisa Cartwright, Constance Penley, and Janet Walker. Nicole Starosielski would also like to thank Jeff Scheible, Lily Chumley, Ben Kafka, Erica Robles-Anderson, Jamie Skye Bianco, and Chi-hui Yang. She is especially grateful to Lisa Parks for her inspiration and collaboration on all of the Signal Traffic projects. Lisa Parks thanks Jennifer Holt, Cristina Venegas, Amelie Hastie, Moya Luckett, Wendy Chun, Jim Schwoch, and Nicole Starosielski for their brilliant ideas, warm friendship, and support, and conveys special thanks to Starosielski for helping to catalyze *Signal Traffic* and for being a remarkable co-editor. Parks is also grateful for support from the Institute for Advanced Study (Wissenschaftskolleg) in Berlin and the Annenberg School of Communication at the University of Pennsylvania, where she held visiting positions that enabled her to work on this project. Parks also thanks John Harley for being the best life-support system imaginable, her dog Luna for getting her away from the computer, and her cat Wink for keeping her company while she is there.

Signal Traffic

Introduction

LISA PARKS AND NICOLE STAROSIELSKI

Signal traffic refers to the movement of electronic media across various parts of the planet. It is the aggregate result of a global culture of continuous electronic transmissions. Though electronic signal trafficking can be dated to the rise of telegraphy during the nineteenth century, this book focuses on the contemporary era of media globalization—an era characterized by contradictory global mediascapes and multiple *media infrastructures*.[1] Today, broadcasting, cable, satellite, Internet, and mobile telephone systems are used simultaneously, and sometimes in coordinated ways, to route signal traffic to and from sites around the world. The content and form of contemporary media—whether television programs or online games—are shaped in relation to the properties and locations of these distribution systems.[2] Simply put, our current mediascapes would not exist without our current media infrastructures. As a suggestive concept, then, signal traffic demarcates a critical shift away from the analysis of screened content alone and toward an understanding of how content moves through the world and how this movement affects content's form. The chapters in *Signal Traffic* call attention to the media infrastructures that distribute audiovisual content, the ways industries and people imagine, organize, and use those infrastructures, and the varied scales at which they operate.

Inside this sprawling brick complex in Hamina, Finland (figure I.1), banks of computers process enormous amounts of data. Located in the icy climate of northern Europe, where the cooling of constantly running electronics is more

Figure I.1. A Google data center sits in the icy landscape of Hamina, Finland, where system operation is more energy efficient.

energy efficient, Google paid $350 million Euros to transform this site from a paper mill into "one of the most advanced and efficient data centers in the Google fleet."[3] The facility, which once employed 650 workers to turn wood into paper, now employs one hundred Google workers to route bits through networks.[4] Just one node in Google's expansive global infrastructure, the Hamina data center is used to distribute Internet traffic primarily throughout Europe. The emergence of such data centers in sites around the world evinces a series of changes in infrastructures of media distribution. Beefed-up broadband pipelines, cloud computing systems, digital compression techniques, and protocols are now integral to the movement and storage of audiovisual signals worldwide.

Just as a paper mill can be repurposed as a data center, a massive water tower can double as a cell-phone mast. In this scene in Lusaka, Zambia (figure I.2), another kind of infrastructural archaeology surfaces as antennas that relay mobile phone traffic are mounted on a tower used to distribute water. Space atop the water tower is leased to commercial mobile-phone operators who appropriate the tower's height to circulate signals and display giant billboards within new "footprints" or "coverage zones." This layering of an emergent system upon an existing one not only exposes the path dependencies of infrastructural

the physicality / materiality of the internet

Figure I.2. In Lusaka, Zambia, mobile phone providers lease space on water towers to send signals and advertise their services.

formations but also reveals how an established node can be used to generate new markets and economic potentials. The water tower no longer only distributes water: it develops a "second life" by hosting a mobile phone tower. As mobile-phone infrastructure is bundled with water infrastructure, sociocultural and economic activities around this node have the potential to alter and expand. In other places around the world, too, mobile-telephone towers have been propped upon skyscrapers, church steeples, minarets, or giant standalone poles; they are sometimes even camouflaged as trees.[5] Built environments have been transformed into wireless footprints. Media and communication researchers have begun to explore the sociocultural and economic relations of mobile telephony, but few have considered the complex materialities of its infrastructure.

Finally, this photograph (figure I.3) features the landing station of the first telephone cable across the Pacific, a link that solidified Hawai'i's position as a communications hub during the 1950s and 1960s. Today, fiber-optic cables extend this legacy, shuttling mobile telephone conversations and Internet traffic across O'ahu's shores. This first cable station at Hanauma Bay, buried

Figure I.3. At Hanauma Bay, Hawai'i, a Cold War–era undersea cable station was buried underground in order to protect it from attack.

underground like many Cold War–era infrastructures, was disguised as part of the hillside to protect it from attack.[6] The burying of this station was not its only environmental impact: installers had to dynamite their way through a reef to ensure the cable had a safe path out to sea. More than fifty years later, this path has become a corridor of environmental tourism. The hole in the reef, now dubbed the "Telephone Cable Channel," draws scuba divers and snorkelers through one of Hawai'i's foremost nature preserves. The cable, once used to traffic telephone calls, has been repurposed by marine scientists to monitor the undersea environment, sensing aquatic life forms and seismic movements on the ocean floor. Critical studies of such sites draw attention to media infrastructures' entanglements with environmental and geopolitical conditions, from the moment of installation through their residual uses.[7]

In this book we conceptualize sites such as data centers, mobile-telephone towers, and undersea cables as *media infrastructures*—situated sociotechnical systems that are designed and configured to support the distribution of audiovisual signal traffic. Media infrastructures are concentrated in particular locations and spread across vast distances. They are highly automated, relying on sensors and remote control, and require human labor for their design, installation, maintenance, and operation. They operate ethereally, transmitting signals at the speed of light, and are grounded in bunker-like facilities heavily secured on earth. Media infrastructures are material forms as well as discursive constructions. They are owned by public entities and private companies and are the products of design schemes, regulatory policies, collective imaginaries, and repetitive use. Interwoven within political-economic agendas, media infrastructures have historically been used in efforts to claim and reorganize territories and temporal relations.[8] Their material dependence on lands, raw materials, and energy imbricates them within issues of finance, urban planning, and natural-resource development.

What can media and communication studies gain by adopting an *infrastructural disposition*? First, a focus on infrastructure foregrounds *processes of distribution* that have taken a backseat in humanities-based research on media culture, which until recently has tended to prioritize processes of production and consumption, encoding and decoding, and textual interpretation.[9] In humanistic media studies there is a serious disjuncture between the amount of scholarly attention dedicated to screened entertainment and the amount devoted to understanding the infrastructures that distribute the signals that become entertainment, whether they exist under the sea, across lands, or "in the cloud."[10] Beyond a concern with the physical systems of media distribution, critical analysis of infrastructure involves interrogating the standards and formats necessary to route content across these systems, whether compression technologies or Internet protocols.

Second, a focus on infrastructure brings into relief the unique *materialities* of media distribution—the resources, technologies, labor, and relations that are required to shape, energize, and sustain the distribution of audiovisual signal traffic on global, national, and local scales. Infrastructures encompass hardware and software, spectacular installations and imperceptible processes, synthetic objects and human personnel, rural and urban environments. Drawing from work in new materialisms and feminist science and technology studies, media infrastructure studies set out to understand the materialities of things, sites, people, and processes that locate media distribution within systems of power.[11] As Diana Coole and Samantha Frost write, "Materiality is always something more than 'mere' matter: an excess, force, vitality, relationality, or difference that renders matter active, self-creative, productive, unproductive."[12] Using a combination of discursive, archaeological, phenomenological, and ethnographic approaches, *Signal Traffic*'s contributors investigate the complex materialisms of infrastructure in a range of locations, from architectural designs in New York City to cybercafés in Turkey, from mobile phone networks in the Middle East to undersea cables in the Pacific. By exploring material forms and practices across national contexts, their chapters bring new settings, objects, and stakeholders into the arena of media and communication research.

Third, a focus on infrastructure compels critical assessment of the relation between *technological literacies* and public involvement in infrastructure development, regulation, and use. Arguably, one of the reasons that infrastructures and "public utilities" have been so steadily privatized by governments over the past several decades is a lack of citizen knowledge about and interest in such systems. As scholars have observed, infrastructures are defined by their invisibility: most of us hardly notice them until they fail or break down.[13] Public access to technical knowledge about infrastructures is not equal; rather, it is guided and constrained by social hierarchies of gender, race/ethnicity, class, generation, and nation. Capitalist societies generally educate people to appreciate the "conveniences" and "choices" of modern consumer technologies, but to remain blind to the infrastructures that support them. As a result, infrastructural changes often occur quickly and without notice, short-circuiting citizen-users' ability to participate in system development. What would it take to arouse greater public interest in media infrastructures? What kinds of scholarship and teaching would help to catalyze and sustain broader citizen involvement in infrastructural matters? It is our hope that the critical study of media infrastructures will deepen scholarly and public engagement with such questions.

Paths to Media Infrastructure Studies

The kinds of systems we define as media infrastructures have historically been referred to by media and communication scholars as *telecommunication networks*. Key research on networks from the telegraph to the Internet have been penned by Harold Innis, James Carey, Herbert Schiller, Benedict Anderson, Armand Mattelart, Manuel Castells, Monroe Price, Jill Hills, and Dan Schiller, among others.[14] Collectively, their scholarship has described the political and economic strategies and regulatory structures that undergird the development of national and international telecommunication systems, the cultural impacts of their emergence, and the imperializing dimensions of their use. This work has articulated the rise of telecommunication networks with the administrative maneuvers of states, governmental agencies and multinational corporations, processes of modernization, urbanization, and globalization, and various stages and forms of capitalism. In other fields, such as history, science and technology studies, geography, and anthropology, scholars have approached infrastructures as large technical systems, urbanization campaigns, and sites of material culture. This interdisciplinary scholarship, which we call *critical infrastructure studies*, draws upon methodologies and frameworks across the humanities and social sciences to historicize and analyze infrastructures ranging from bridges to power grids, from railways to sewer systems.[15]

Building upon this research, we understand media infrastructures not only as telecommunication networks owned and operated by governments, militaries, and corporations, but as complex material formations that operate at multiple *scales*. We describe these formations using a *relational approach* that recognizes the industrial, physical, and organizational interconnections of media infrastructures with other systems. We address the *different and uneven conditions* that shape and characterize media infrastructures around the world as well as the *labor, maintenance, and repair* required to build and sustain them. Our approach also considers the natural resources that media infrastructures require and the environmental impacts they produce. We further attend to the myriad ways people encounter, perceive, and use media infrastructure—that is, the *affective relations* they generate and become part of. Finally, critical studies of media infrastructures, we believe, can provide a platform for *innovative methodologies* by activating and combining approaches such as archaeology, political economy, phenomenology, ethnography, and discourse analysis. In the sections that follow, we further discuss these critical issues and some of the research that informs them.

Scale

One of the most distinctive aspects of media infrastructures is their scale: they span continents, oceans, and atmospheres, and can leave long-lasting traces. Some work in critical infrastructure studies foregrounds the significance of scale by documenting the relations between large technical systems and processes of industrialization. In his influential book *Networks of Power*, Thomas Parke Hughes uses the case of electrical systems to extrapolate several phases of infrastructure formation, including invention and development, transfer between regions and societies, system growth, and the attainment of technological momentum.[16] By establishing a general framework for studying infrastructures as large technical systems, Hughes inspired histories of other such systems, including railroads, telecommunication, air-traffic control, and gas networks.[17] Historians of nineteenth-century culture and technology, for instance, have described how large networks of transportation *the cohabitat of multiple scales*

and communication "annihilated" space and time, facilitated standardization, and reshaped everyday life. Building on the work of Hughes and others, Paul Edwards argues that large-scale infrastructures are core to the experience of modernity, observing, "To be modern is to live within and by means of infrastructures."[18] Yet Edwards insists that mesoscale studies of infrastructure, such as Hughes's, tend to generalize about and normalize conditions of modernity, failing to account for the fact that people often "inhabit, uneasily, the intersection of . . . multiple scales."[19] *Signal Traffic* heeds Edwards's call for more macroscale and microscale studies that explore a broader range of national and user contexts and attend to variable infrastructural conditions.

Approaching infrastructure across different scales involves shifting away from thinking about infrastructures solely as centrally organized, large-scale technical systems and recognizing them as part of multivalent sociotechnical relations. Rather than take an overarching or mesoscale view, digital media and informatics researchers have honed in on the macroscale and microscale elements of networks, protocols, and bits, investigating material-semiotic and experiential dimensions of digital technologies. Foundational studies by Wendy Chun and Alexander Galloway, for instance, have foregrounded the macrolevel fiber-optic networks and microscale protocols through which data circulate, respectively.[20] Jean-François Blanchette has delved into the nitty-gritty of computing by examining bits, insisting that they "cannot escape the material constraints of the physical devices that manipulate, store and exchange them."[21] These works, among others, have contributed to emergent fields of software studies and platform studies, the latter of which examines the hardware on which software runs and digital media are materialized.[22]

In an effort to recognize the range of scales at which infrastructures operate, the chapters here investigate the dynamic components of media infrastructures in ways that enrich and deepen macroscale and microscale analysis. Building upon Galloway's research, in this book Paul Dourish explores key design issues in the development of Internet protocols and demonstrates how and why the size of data packets traversing through networks matters. Decisions about whether a message should be broken into 64-byte or 32-byte "payloads," he reveals, are related to the divergent characteristics of national infrastructures and geographies. Jonathan Sterne's chapter similarly moves between scales, charting the historical emergence of microscale compression technologies in relation to macroscale transmission lines. As these chapters explore how microscale processes and macroscale architectures inflect one another, they bring the complex materialities and relationalities of media infrastructures into focus.

Relationality

In addition to recognizing the multiscalar dimensions of infrastructures, *Signal Traffic*'s contributors emphasize the layering or bundling of distinct systems (such as that of water and mobile telephony discussed earlier) as well as the interconnections between infrastructures, environments, and users. Researchers in science and technology studies approach infrastructures as dynamic sociotechnical formations and organizations rather than as isolated or static machines. According to Susan Leigh Star and Karen Ruhleder, infrastructure encompasses both technical bases and social arrangements, extends beyond single events and sites, connects with existing practices and standards, and must be learned and naturalized over time by users.[23] As such, infrastructure is fundamentally a relational concept rather than a concrete object; it "is something that emerges for people in practice, [and is] connected to activities and structures."[24] Blending approaches from sociology, communication, and anthropology, Susan Leigh Star and Geoffrey Bowker have revealed how infrastructures are embedded in everyday practice, foregrounding the hidden labor they rely upon as well as how they are contingent on social structures. For them, infrastructure refers not only to tubes and pipes but includes "soft" systems of organization and knowledge, ranging from professional societies to classificatory procedures. Infrastructure studies, their work demonstrates, is not simply a quest to understand large technical systems; rather, it explores processes and changes at a "mundane scale" and treats them as part of the building of organizations and production of knowledge.[25]

Some research on infrastructure, including that of Bowker and Star, builds upon and extends Actor-Network Theory (ANT), developed by Michael Callon, John Law, and Bruno Latour.[26] ANT insists on the complex relationalities of social and technical systems, and it troubles the tendency to reduce or ignore the agential aspects of nonhuman objects as well as the responsibilities that humans delegate to them.[27] Researchers in the areas of organizational communication and informatics have drawn upon ANT to create an interdisciplinary field known as *information infrastructure*. Work in this field has set out to rethink the ontology of infrastructures, critiquing assumptions of their stability and manageability, and treating infrastructures as "performative forces that evolve dynamically" and as phenomena that are "generated and regenerated in open-ended relationships."[28]

Other theorizations of relationality have emerged in recent work on "new materialisms," which, like ANT, emphasizes complex relationships between human and nonhuman actants. Karen Barad's reconceptualization of materiality, for example, tasks us to see the material world not as simply given or independent but as ontologically entangled with and produced through the very apparatuses we use to make sense of it.[29] Inspired by Barad, Coole and Frost argue that we must think about causation in much more complex ways and "recognize that phenomena are caught in a multitude of interlocking systems and forces and to consider anew the location and nature of capacities for agency."[30] Objects have a life, according to vital materialist Jane Bennett, because of their capacity to make a difference in the world and to have effects. Approaching the power grid in a very different way than Thomas Hughes, Bennett conceptualizes it as an assemblage or "federation of actants," explaining: "To the vital materialist the electrical grid is . . . understood as a volatile mix of coal, sweat, electromagnetic fields, computer programs, electron streams, profit motives, heat, lifestyles, nuclear fuel, plastic, fantasies of mastery, static, legislation, water, economic theory, wire, and wood—to name just some of the actants."[31] After uncovering the litany of causes and effects of power outages and blackouts in North America, she insists, "humans are not the sole or most profound actants in assemblages."[32]

Feminist critics of science and technology such as Donna Haraway, Rosi Braidotti, Lucy Suchman, and Chela Sandoval have for decades been insisting upon the need for ontologies and epistemologies that recognize a broader and more diverse spectrum of human/nonhuman hybrids, interactions, and relations. Emerging research in ANT, new/vital materialisms, and object-oriented ontologies reveals that a broader intellectual quorum has formed around the idea that objects matter.[33] Although recent work in the areas of object-oriented ontologies, media archaeology, and platform studies addresses the materiality of technological systems, much of it overlooks feminist critiques of technology, power, and difference that are integral to our conceptualization of media infrastructures. Tarleton Gillespie's astute interrogation of the "politics of platforms," however, serves as an exception, as it confronts the ways power and discourse help to constitute what a "platform" is and who controls it.[34] Though only some of the chapters in this book are influenced by research in these emergent fields, we anticipate that future research on media infrastructures will engage more directly with this work as it challenges us to recognize a more extensive field of actants and relations in media and communication studies. Authors in this collection, for instance, show that in some parts of the world Internet and mobile-phone infrastructure could not function without water, state surveillance could not occur without land and spectrum, and data centers could not function without the sun. In such scenarios, humans are but one part of broader infrastructural formations.

Difference and Unevenness

As *Signal Traffic*'s contributors examine media infrastructures across scales and as complex relationalities, they also explore how these extensive systems emerge in different parts of the world. What is often missing from mesoscale accounts is a detailed investigation of the varied ways that infrastructures intersect with cultures of everyday life as well as how their implementation and use fluctuates across industrialized and developing regions, rich and poor neighborhoods, and urban and rural settings. Media infrastructures may be centrally owned by nation-states or corporations, but at their edges they are imagined, arranged, and adopted in different ways by people or "end-users." As Colin McFarlane and Jonathan Rutherford have argued, we must provincialize the study of infrastructure and examine how it matters differently to various groups across space and time.[35] Toward that end, some chapters of this collection explore what media infrastructures look like or feel like from a peoples' or populist perspective.[36] Lisa Parks's chapter, for instance, considers how people's use of the Internet in rural Zambia is punctuated by variable access to electricity and water. Helga Tawil-Souri's chapter explores how Palestinians experience the political topography of the occupied territories in their encounters with mobile telephony.

In an effort to highlight the differential dimensions of infrastructures, urban studies scholars have conducted fieldwork in cities around the world. In their pathbreaking book, *Splintering Urbanism*, Stephen Graham and Simon Marvin demonstrate how networked infrastructures across sectors of energy, telecommunication, transportation, and water have been organized in ways that support the privatization of public utilities and create urban fragmentation.[37] Immersing readers in specific infrastructure nodes in cities north and south, Graham and Marvin challenge us to develop site-specific investigations of the "massive technical systems that interlace, infuse and underpin cities and urban life" and to participate in the politics of their future imagining and formation.[38] Their work offers a crucial model for studying infrastructures across global/national/local contexts, in relational ways, and in close-up (in situ), and informs much of the research in this book. Extending the focus beyond urban settings, *Signal Traffic*'s contributors offer studies of rural or transitional areas, bringing new ecologies, technological objects, and communities into infrastructure research.[39]

As infrastructures emerge in relation to conditions of difference and unevenness, they are fraught within relationships of power. The organization and use of infrastructures have the potential to reinforce or reverse unjust social relations. Insisting upon the need to address the politics of digital networks, scholars such rk Poster and Manuel Castells have argued that the Internet augured a new "mode of information"[40] or "space of flows"[41] permeated by power differentials. Corresponding research on "cyberinfrastructure" has emphasized the levels of technological access and literacy that digital systems require and the new divides they can create.[42] Drilling down on this point, researchers have also confronted the politics of the "digital divide," explicating how and why access to media infrastructures relates to disenfranchisement and exclusion.[43] This question of who has access to digital technologies arguably remains one of the most pressing issues of our times, and an entire field called Information and Communications Technologies for Development (ICTD) has emerged to try and tackle it.

Even with infrastructures in place and broadly accessible, there is no guarantee that they will function properly or serve people's interests. As anthropologists Dennis Rogers and Bruce O'Neill point out, infrastructures also can have deleterious effects, enforcing social norms or enacting physical and emotional harm.[44] Rogers and O'Neill argue that in certain situations "infrastructure is not just a material embodiment of violence (structural or otherwise), but often its instrumental medium."[45] More than simply to divide people, infrastructures can be used to exert force or injure. Turning off the electrical grid during times of war means civilians freeze. Making telecommunication costly can shut the poor off from emergency services and put lives in jeopardy. The shift from a normalized condition of infrastructural service and connection to one of disruption and disconnection, whether because of war, weather, or cost, can create profound physical and psychic experiences for communities and individuals alike.

Labor/Repair/Maintenance

Studies of media infrastructure also must take into account labor, maintenance, and repair, since system operations depend on these practices. As Nigel Thrift suggests, an infrastructure must be produced and reproduced through social practices: "[it] has precisely to be performative, if it is to become reliably repetitive."[46] Research by Carolyn Marvin, Greg Downey, and Brian Larkin has addressed the performative labor of infrastructure by exploring, respectively, electricians' imaginings of early power grids, the dynamic movements of messenger boys who fueled early telegraphy systems, and the colonialists who built bridges and radio networks to extend their ways of life into Africa.[47] Sarah Harris's chapter builds upon this research by demonstrating how cybercafé operators' daily routines become part of Internet infrastructure in Turkey. Turkish cybercafé operators are able to maintain Internet connectivity, Harris suggests, only by constantly renegotiating and selectively enforcing state censorship policies in their neighborhood shops.

The operation of media infrastructures is contingent not only on the labor of those who operate or maintain them on a daily basis but also on those who build end-devices, whether smart phones, laptops, or high-definition TVs. In *Below the Line*, Vicki Mayer uncovers the life worlds of TV-set manufacturers in Brazil, revealing that it is impossible to separate the global distribution of entertainment media from those who spend tireless hours on assembly lines manufacturing the electronic devices used to consume it. Charles Acland's chapter in this book foregrounds the fact that consumer electronics are vital to the transmedia era and suggests that they support the "platform plenitude" and "branded viewing experiences" of a "wired class." Combined, Mayer's and Acland's work suggests the need for further research on the manual and intellectual labor and industrial conditions upon which media infrastructures are built.

Given the growing economic investment in and cultural fascination with audiovisual infrastructures, platforms, and devices, it is important to consider what happens when such systems malfunction or fail.[48] In *Disrupted Cities: When Infrastructure Fails,* Stephen Graham argues that studying moments of breakdown or failure might be the most appropriate heuristic device for infrastructural understanding, for it is "perhaps the most powerful way of really penetrating and problematizing those very normalities of flow and circulation."[49] Moments of failure, in other words, can help to reveal or bring into consciousness the myriad micro- and macro-level conditions and perceptions of "flow and circulation" that are needed to sustain infrastructural operations in the first place. Consistent with this contention, Steven Jackson suggests an epistemic shift toward what he calls "broken world thinking," asserting that "breakdown, dissolution, and change, rather than innovation, development, or design . . . are the key themes and problems facing new media and technology scholarship today."[50] Research in this area suggests that infrastructural breakdowns and acts of repair should be thought about as a "normal" part of technological processes and as opportunities for retooling social relations.[51]

Natural Resources/Environment

Another of this book's critical interventions is to focus further attention on the relationship between media infrastructures, natural resources, and environments. As work by Harold Innis and James Carey has shown, the organization and physical arrangement of media infrastructures demand critical thinking across sectors of energy, transportation, agriculture, natural resources, and trade. Richard Maxwell and Toby Miller's *Greening the Media* details the heavy resource demands and environmental impacts of the contemporary global media economy: they report that in 2007 media technologies were responsible for between 2.5 percent and 3 percent of the world's greenhouse gas emissions, a figure that has only increased with the expansion of Internet infrastructure, emergence of new data centers, and intensified production and use of consumer electronics.[52] As Nadia Bozak argues in her important book *The Cinematic Footprint*, "The image—cinematic, photographic, digital, or analog—is . . . materially and economically inseparable from the biophysical environment."[53] We would add to Bozak's claim that image (and sound) distribution—signal trafficking—is also inseparable from the biophysical environment. *Signal Traffic*'s contributors not only consider the resource requirements of media infrastructures, they also explore how the availability of water, land, electricity, and spectrum can determine the geographic positioning and physical organization of infrastructures such as transoceanic cables, networked data centers, and mobile-phone towers.

In this way, the critical study of media infrastructures is tied directly to the emergent field of *environmental media studies* as it considers where the materials and energy needed to build, operate, and sustain massive systems of content distribution come from and evaluates the impacts of those systems on environs in different parts of the world.[54] In pursuing such issues, research in this book also sets out to complicate epistemological divides between technology and nature, human and nonhuman, material and immaterial, suggesting that such categories are relationally defined and materially intertwined. By emphasizing the entanglement of media infrastructures and environments, this book embraces Sarah Kember and Joanna Zylinska's provocative suggestion that "mediation can be seen as another term for 'life,' for being-in and emerging-with the world."[55] In practice, this approach troubles any clear distinction between what we consider to be media infrastructure, such as a broadcast transmitter, and sites and processes typically thought of as its "environment." Infrastructures and environments dynamically mediate and remediate one another. Ashley Carse argues that as natural environments are increasingly shaped by human action, phenomena such as rivers and forests have been transformed into systems of human imagination and intervention, rendering nature itself infrastructural.[56] This shift raises questions such as: How do the rains in rural Zambia or the rivers in Oregon become infrastructural for media circulation? What kinds of media distribution do these "natural" environments support? How are nonhuman forms of life affected by the presence of media infrastructures?

~ not a horrific

Affect

In addition to exploring the relationship between media infrastructures, natural resources, and the environment, infrastructure can be studied as part of an "affective turn."[57] In their introduction to *The Affect Theory Reader*, Greg Siegworth and Melissa Gregg point out that "affect" has a complex history with many valences. Drawing upon phenomenological philosophies, they define affect generally as "a gradient of bodily capacity—a supple incrementalism of ever-modulating force-relations—that rises and falls not only along various rhythms and modalities of encounter but also through the troughs and sieves of sensation and sensibility."[58] To be sure, infrastructures are part of such "force-relations," since our encounters with them can elicit different dispositions, rhythms, structures of feeling, moods, and sensations. For many people, the default affective response to infrastructure might be apathy, disinterest, or indifference, but it is also possible that a broad spectrum of infrastructure-related affects remains unspoken and unknown simply because certain questions have not been asked.

Darin Barney's ethnographic study of "grain-handling technologies" and "railway branchlines" on the Canadian prairies is an exemplary study of infrastructure and affect. Immersing himself in the life worlds of small-town grain farmers, Barney describes grain silos and railroads as places of focused attention and exchange in rural communities.[59] One of his informants describes the grain elevator at Fairlight, Saskatchewan, as "a place to hear the news—news of births and deaths and war and peace. It's been a place to debate politics, wheat prices, wheat boards and hockey; a place to shake the loneliness of life on the land."[60] The takeover of these facilities by big agribusiness during the past two decades, Barney explains, not only resulted in the gradual demolition and replacement of these infrastructure sites with more "efficient" farming equipment, but the shift also generated feelings of isolation and frustration as farmers sat in long lines alone in their trucks waiting to unload grain in conglomerates' new "through-put terminals." By shedding light on "the complex ways in which infrastructural technologies mediate the organization of social and political life," Barney's research brings affective dimensions of infrastructures to the surface, while bringing different objects and actants into the repertoire of media studies.[61]

A phenomenology of infrastructure and affect might begin by excavating the various dispositions, feelings, moods, or sensations people experience during encounters with infrastructural objects, sites, and processes. This exercise could unfold along a continuum that recognizes, on one end, the general tendency of infrastructures to normalize behavior (such that they become relatively invisible, unnoticed, or internalized), and, on the other, as the potential for the disruption of that normalization, which can occur during instances of inaccessibility, breakdown, replacement, or reinvention. By sketching out this continuum, we build upon Wendy Chun's crucial work on the Internet's relation to control and freedom and point to the cornucopia of infrastructural affects that lies in the gray zone between them. We hope that this will catalyze further thinking about the range of ways people perceive and experience infrastructures in everyday life and how these experiences differentially orient or position people in the world.

Innovative Methodologies

Finally, in addition to approaching media infrastructure as a site for critical thinking about issues of scale, relationality, difference and unevenness, labor and maintenance, natural resources, and affect, we think of the concept as a nesting ground for innovative research methodologies. As Bowker, et al. have argued, "Infrastructure studies require drawing together methods that are equal to the ambitions of its phenomenon."[62] And as Brian Larkin suggests, "The sheer diversity of ways to conceive of and analyze infrastructures . . . cumulatively point[s] to the productive instability of the basic unit of research."[63] Like infrastructures, research units and methods are dynamic fields that take time to emerge and solidify. *Signal Traffic* brings together projects that use qualitative methodologies such as discourse analysis, ethnography, archaeology, archival research, industry analysis, and fieldwork. As multidisciplinary scholars situated primarily in the humanities, our contributors also bring a range of critical theories to bear on the study of media infrastructures, drawing from poststructuralist theories of power, postcolonial criticism, science and technology studies, feminist theory, historiography, and cultural geography. The result is a broad tapestry of approaches. While some chapters delineate the conceptualization and historicization of infrastructural processes, others examine specific infrastructure sites or objects. While some focus on centers of infrastructural activity, others explore infrastructural edges, outskirts, or fringes as well as those Susan Leigh Star once referred to as the "orphans of infrastructure."[64] And while some focus on imperceptible, microscale phenomena, others take a step back and provide a big picture. Since media infrastructures are configured in relation to and sometimes literally built on top of other infrastructures, they also invite archaeological approaches. In her chapter of this book Shannon Mattern engages with such approaches to conceptualize what she calls "the deep time of media infrastructure."

Collectively, the work in *Signal Traffic* sets out to extend materialist studies of media technologies by rethinking and expanding the concept of infrastructure, exploring physical installations, objects, sites, and processes in detail, analyzing industrial transitions, and probing the sociohistorical conditions and power relations that give shape to particular infrastructural formations. Contributors approach the global mediascape as a contradictory and contested domain that must be engaged in multiple ways, from historical, political economic, and sociotechnical perspectives. They explore media infrastructures from the top down and the bottom up, in urban and rural space, and in high- and low-tech conditions. They are mindful of blockages as well as flows, and pay attention to the intersections of meso, macro, and micro scales and processes. The book features field-based ethnographies and archival research alongside studies of industrial forces, technical design, and labor. It explores contemporary media infrastructures such as the Internet and mobile phone networks in relation to water systems, solar power, and human energy. And as the book traces the emergence of infrastructural hardware and installations, it also includes discussions of "soft" infrastructures such as daily routines, marketing, and knowledge practices. *Signal Traffic* engages with media infrastructure as a concept and material formation, positions it in relation to the politics of difference, and tracks it across different parts of the world, from Sweden to Palestine, from Turkey to Zambia.

blockles in a per day TOYS
time-space compression

The Collection

Signal Traffic is organized into three parts. The first, "Compression, Storage, Distribution," features historical and contemporary conceptualizations of media infrastructures as well as analyses of the changing capacities to format, store, and distribute media, whether on disks, through cables, or in clouds. The section opens with Jonathan Sterne's genealogy of media compression techniques and their relation to infrastructures that have historically been developed and scaled to carry or transmit certain loads or capacities. Sterne suggests that by examining compression—a process that accommodates signals to infrastructures—it is possible to rethink and rewrite media history away from a general history of verisimilitude and toward a general history of compression. This historiographic intervention might turn further attention to experiences and aesthetics that emerge around media in limited definition. It might also facilitate an understanding of the ways that compression both renders representation adequate to infrastructure and exposes the limits of transmission. In the end, Sterne observes, compression techniques also work upon infrastructures, making them adequate to the representational loads that pass through them. Using examples ranging from audio compressors to the optical telegraph, Sterne demonstrates that content and infrastructure exist in relations of "circular causality."

Also exploring the relationship between media content and the capacity of hard infrastructure, Nicole Starosielski's chapter, "Fixed Flow: Undersea Cables as Media Infrastructure," offers a framework for understanding how particular technologies, social practices, and natural environments can be conceptualized as media infrastructures. Drawing from work by Susan Leigh Star and Karen Ruhleder, she develops a relational approach to media infrastructure that delineates the multiple routes and effects of global undersea cable networks. Her chapter describes five of the ways undersea cables function as a media infrastructure: they become resources for media activity; alter our everyday experience of media temporality; shape our susceptibility to media censorship and surveillance; solidify global relationships of media power; and serve as a platform where publics can affect the dissemination of media content.

Shifting the focus from transoceanic cables to data centers and cloud computing, Jennifer Holt and Patrick Vonderau's chapter explores how recent depictions of data-center visibility function both as a mode of claiming corporate territory and as an obfuscation of the less picturesque dimensions of cloud infrastructure. As Holt and Vonderau excavate the material support systems, standards, protocols, and constraints of cloud computing, they suggest that analyzing media infrastructure industries, such as the companies that run cloud systems, presents particular challenges for researchers. According to Holt and Vonderau, the structural convergence and functional heterogeneity of media make it difficult to apply some of the tried and true concepts in media and communication studies, such as the distinction between public and private. Using the Swedish data center as an example, Holt and Vonderau decipher the backend of Internet architecture and data-trafficking policies, and they highlight the importance of a relational perspective in understanding data centers as dynamic infrastructure nodes.

In "Deep Time of Media Infrastructure" Shannon Mattern establishes the significance of historical media infrastructures that precede the digital era. Adopting a media archaeological approach, Mattern explores how historical networks layered in urban space shape contemporary media systems. These networks extend back far beyond nineteenth-century telegraph wires to include much earlier Greek-inspired aural, inscriptive, and architectural forms. Suggesting that research on early media infrastructures can usefully inform studies of the media city, which typically begin with modern media and rarely include

a new techo-archedomy JM/
infrostru is TN -
Introduction • 19

TN detorall stop support

discussions of infrastructure, Mattern delineates a number of potential inter-disciplinary engagements for media infrastructure studies, ranging from geol-ogy to architectural history. Her chapter closes with an important discussion of what media studies can gain from further engagement with archaeological and infrastructural research.

The book's second part, "Resources, Environments, Geopolitics," features a series of site-specific case studies that explore how different configurations of energy, territory, state power, and local practices affect the shape and form of infrastructures as well as knowledge about and access to them. The part begins with Lisa Parks's chapter "Water, Energy, Access: Materializing the Internet in Rural Zambia." Drawing on ethnographic fieldwork, Parks describes a particular rural configuration of Internet infrastructure and shows that access in this loca-tion is contingent on water resources, which not only generate hydroelectricity for the Zambian power grid but are also necessary for prospective Internet users' everyday survival in the community of Macha, Zambia. Her chapter foregrounds the struggles and contestations that are part of infrastructure development; the energy and biopower that infrastructures rely on; the relationality of water, transportation, and information systems; and the alternate ways that people imagine, use, or respond to infrastructure, which may range from intense cu-riosity to patent disinterest.

Also concerned with the topic of energy, Toby Miller's chapter, "The Art of Waste: Contemporary Culture and Unsustainable Energy Use," provocatively challenges media and cultural studies to confront the environmental impacts of the global digital economy. After critiquing an array of intellectual and cor-porate discourses that celebrate the beneficence of digital technologies, Miller proposes what he calls the "art of waste" and brings a discussion of e-waste together with critiques of the art of labor and the cognitariat. As he insists, "rather than seeing new communication technologies as magical agents that can produce market equilibrium and hence individual and collective happiness, we should note their other impacts." The chapter concludes with a discussion of e-waste–related art projects, which, Miller argues, have the capacity "to exemplify and criticize a state of affairs that must not be allowed to continue." Miller's chapter thus addresses macrolevel environmental and resource ques-tions that underpin the critical study of media infrastructures.

Weaving geopolitics into this part's discussion of energy resources and media infrastructures, Helga Tawil-Souri's chapter details the conditions and contestations underlying cellular phone infrastructures in Israel-Palestine. As she shows how cellular infrastructures in the occupied territories are dynamic manifestations of territorial disputes and tensions, Tawil-Souri argues that the arrangement of telecommunication systems is not merely a metaphor for the conflict; rather, "it is the conflict in material form." Her chapter focuses on three locations—Migron, Ramallah, and Qalandia—and describes the material infrastructures and regulatory regimes that shape conditions in each. Rather than connecting people, she argues, these infrastructures are critical dimensions of state power and territoriality, and as such they function in ways that divide and disconnect.

The book's third part, "Content, Protocols, Platforms," opens with Paul Dourish's meticulous analysis of the materialities of Internet protocols. Returning to some of the issues addressed in Sterne's chapter, Dourish focuses on the relationship between content and conduit, which involves both the compression and modulation of signals. Dourish argues that we need to look not only at the materialities of hard infrastructural elements—from buildings to antennae—but also at the materialities of protocols themselves. He directs attention to the relationships between infrastructures and experience, and the micro-level processes by which digital experiences are produced. To address these concerns, Dourish details the development of Internet routing protocols, tracing how they tie networks together and mediate between hard infrastructure and the circulation of content. He contrasts two different protocols, the Routing Information Protocol and the Exterior Gateway Protocol, which emerged in different historical moments and cultural conditions. Examining the social construction of these network protocols, he reminds us, can help us to differentiate the actual Internet—which grows out of specific material constraints—from a possible or imagined Internet.

Also concerned with the issue of Internet protocols, Sarah Harris's chapter, "Service Providers as Digital Media Infrastructure: Turkey's Cybercafé Operators," approaches the topic in a different manner, focusing on circumvention practices in Turkey. Building upon the literature on infrastructural labor, Harris documents the critical role of service providers in the development of today's digital media systems. She illustrates how an ethnographic approach to media infrastructures helps to connect hard infrastructural forms, such as wires, transmissions towers, and buildings, with soft infrastructural forms, including institutions, protocols, and social practices. Harris suggests that the work of Turkey's cybercafé operators forms a key component of Internet infrastructure, critically shaping the social topography of media in the country. The cafés and their operators coordinate disparate technologies and communities and are sites where different protocols are negotiated. At the same time, Harris shows, in these locations state infrastructural control, surveillance, and censorship can be undermined.

Also delving into particular protocols and platforms, Christian Sandvig's chapter, "The Internet as the Anti-Television: Distribution Infrastructure as Culture and Power," investigates the architecture used to distribute video over the Internet. Noting the unprecedented volume of online video that now circulates, Sandvig suggests that this distribution has "enabled a radical approach" by generating forms of labor and content that traditional media industries have never seen before. Suggesting that "television and Internet traffic were at first like oil and water," he explores how computer pioneers thought about television in the 1960s and charts a path to more recent practices of caching, streaming, and multicasting. The case of Internet video distribution, he argues, reveals how crucial the study of infrastructure is to understanding the shape, form, and function of media technologies.

Concluding the book, Charles R. Acland's chapter, "Consumer Electronics and the Building of an Entertainment Infrastructure," shifts the discussion away from Internet protocols and describes an emergent constellation of protocols and platforms within contemporary Hollywood. Returning to issues raised by Holt and Vonderau and Starosielski in the book's first part, Acland's chapter explores how Hollywood's "technological tentpoles"—films that strategically promote cross-media commodities and new generations of devices, platforms, and hardware—serve as vehicles for the advancement of a broader technological system. As Acland puts it, a "dispersed network of devices forms an entertainment and informational infrastructure upon which dominant cultural and economic practices transpire." Moving between entertainment industry events and a proliferating field of consumer electronics, Acland shows how audiovisual infrastructure is a product not only of economic priorities, but also of the conceptual frames that are circulated about them.

Notes

1. Arjun Appadurai, *Modernity at Large: Cultural Dimensions of Globalization* (Minneapolis: University of Minnesota Press, 1996).

2. Raymond Williams, *Television: Technology and Cultural Form*, 3rd ed. (London: Routledge, 2003).

3. Google, "Hamina, Finland." Available at http://www.google.com/about/datacenters/inside/locations/hamina (accessed September 18, 2014).

4. Steven Levy, "Where Servers Meet Saunas: A Visit to Google's Finland Data Center," *Wired*, October 24, 2012.

5. Lisa Parks, "Around the Antenna Tree: The Politics of Infrastructural Visibility," *Flow* (March 2009). Available at http://flowtv.org/2010/03/flow-favorites-around-the-antenna-tree-the-politics-of-infrastructural-visibilitylisa-parks-uc-santa-barbara (accessed October 14, 2014).

6. Tom Vanderbilt, *Survival City: Adventures among the Ruins of Atomic America* (New York: Princeton Architectural, 2002).

7. Nicole Starosielski, *The Undersea Network* (Durham, N.C.: Duke University Press, forthcoming 2015).

8. Harold Innis, *The Bias of Communication* (Toronto: University of Toronto Press, 1951); James Carey, *Communication as Culture* (New York: Routledge, 1989).

9. Exceptions to this include Sean Cubbitt, "Distribution and Media Flows," *Cultural Politics* 1, no. 2 (2005): 193–215; Jennifer Holt, *Empires of Entertainment: Media Industries and the Politics of Deregulation, 1980–1996* (New Brunswick, N.J.: Rutgers University Press, 2011); Jennifer Holt and Kevin Sanson, *Connected Viewing: Selling, Streaming, and Sharing Media in the Digital Era* (New York: Routledge, 2014); Jonathan Sterne, "Television under Construction: American Television and the Problem of Distribution 1926–1962," *Media, Culture and Society* 21, no. 3 (1999): 503–30; Michael Curtin, "Media Capitals: Cultural Geographies of Global TV," in *Television after TV: Essays on a Medium in Transition*, ed. Lynn Spigel and Jan Olsson (Durham, N.C.: Duke University Press, 2004), 270–302.

10. Lisa Parks, "Earth Observation and Signal Territories: Studying U.S. Broadcast Infrastructure through Historical Network Maps, Google Earth, and Fieldwork," *Canadian Journal of Communication* 38, no. 3 (2013): 19.

11. Diana Coole and Samantha Frost, eds. *New Materialisms: Ontology, Agency, and Politics* (Durham, N.C.: Duke University Press, 2012); Graham Harman, *The Quadruple Object* (Washington, D.C.: Zero, 2011); Jane Bennett, *Vibrant Matter: A Political Ecology of Things* (Durham, N.C.: Duke University Press, 2010).

12. Diana Coole and Samantha Frost, eds., *New Materialisms: Ontology, Agency, and Politics* (Durham, N.C.: Duke University Press, 2012): 9.

13. Susan Leigh Star and Karen Ruhleder, "Steps toward an Ecology of Infrastructure: Design and Access for Large Information Spaces," *Information Systems Research* 7, no. 1 (1996): 111–34.

14. Innis, *Bias of Communication*; Carey, *Communication as Culture*; Herbert Schiller, *Mass Communications and American Empire* (New York: Keeley, 1969); Benedict Anderson, *Imagined Communities: Reflections on the Origin and Spread of Nationalism* (London: Verso, 1983); Manuel Castells, *The Rise of the Network Society* (Cambridge, Mass.: Blackwell, 1996); Monroe Price, *Media and Sovereignty: The Global Information Revolution and Its Challenge to State Power* (Cambridge, Mass.: MIT Press, 2002); Dan Schiller, *Digital Capitalism: Networking the Global Market System* (Cambridge, Mass.: MIT Press, 2000); Armand Mattelart, *Networking the World: 1794–2000* (Minneapolis: University of Minnesota Press, 2000); and Jill Hills, *Telecommunications and Empire* (Urbana: University of Illinois Press, 2007).

15. Christian Sandvig distinguishes different types of infrastructure research, the "relationalist" approach and the "new materialist" approach. Christian Sandvig, "The Internet as Infrastructure," in *The Oxford Handbook of Internet Studies*, ed. William Dutton (Oxford: Oxford University Press, 2013), 86–108.

16. Thomas Parke Hughes, *Networks of Power: Electrification in Western Society, 1880–1930* (Baltimore, Md.: Johns Hopkins University Press, 1983).

17. Wiebe E. Bijker, Thomas Parke Hughes, and Trevor J. Pinch, eds., *The Social Construction of Technological Systems: New Directions in the Sociology and History of Technology* (Cambridge, Mass.: MIT Press, 1989); Renate Mayntz and Thomas Parke Hughes, *The Development of Large Technical Systems* (Boulder, Colo.: Westview, 1988); Oliver Coutard, *The Governance of Large Technical Systems* (New York: Routledge, 1999).

18. Paul N. Edwards, "Infrastructure and Modernity: Force, Time, and Social Organization in the History of Sociotechnical Systems," in *Modernity and Technology*, ed. Thomas J. Misa, Philip Brey, and Andrew Feenberg (Cambridge, Mass.: MIT Press, 2003), 186.

19. Ibid.

20. Alexander R. Galloway, *Protocol: How Control Exists after Decentralization* (Cambridge, Mass.: MIT Press, 2004); Wendy Hui Kyong Chun, *Control and Freedom: Power and Paranoia in the Age of Fiber Optics* (Cambridge, Mass.: MIT Press, 2006).

21. Jean-François Blanchette, "A Material History of Bits," *Journal of the American Society for Information Science and Technology* 62, no. 6 (2011): 1042.

22. Matthew Kirschenbaum, *Mechanisms: New Media and the Forensic Imagination* (Cambridge, Mass.: MIT Press, 2008); Matthew Fuller, ed., *Software Studies: A Lexicon* (Cambridge, Mass.: MIT Press, 2008); Nick Montfort and Ian Bogost, *Racing the Beam: The Atari Video Computer System* (Cambridge, Mass.: MIT Press, 2009); Steven E. Jones and George K. Thiruvathukal, *Codename Revolution: The Nintendo Wii Platform* (Cambridge, Mass.: MIT Press, 2012).

23. Star and Ruhleder, "Ecology of Infrastructure"; Susan Leigh Star, "The Ethnography of Infrastructure," *American Behavioral Scientist* 43, no. 3 (1999): 377–91; Geoffrey C. Bowker and Susan Leigh Star, *Sorting Things Out: Classification and Its Consequences* (Cambridge, Mass.: MIT Press, 1999).

24. Star and Ruhleder, "Ecology of Infrastructure," 112.

25. Geoffrey C. Bowker, Karen Baker, Florence Millerand, and David Ribes, "Toward Information Infrastructure Studies: Ways of Knowing in a Networked Environment," in *International Handbook of Internet Research*, ed. Jeremy Hunsinger, Lisbeth Klastrup, and Matthew Allen (Dordrecht: Springer Science+Business Media, 2010), 111; Geoffrey C. Bowker, *Science on the Run: Information Management and Industrial Geophysics at Schlumberger, 1920–1940* (Cambridge, Mass.: MIT Press, 1994); Bowker and Star, *Sorting Things Out*.

26. Bruno Latour, *Science in Action: How to Follow Scientists and Engineers through Society* (Cambridge, Mass.: Harvard University Press, 1996).

27. Michel Callon, "Society in the Making: The Study of Technology as a Tool for Sociological Analysis," in *The Social Construction of Technological Systems: New Directions in the Sociology and History of Technology*, ed. Wiebe E. Bijker, Thomas P. Hughes, and Trevor J. Pinch (Cambridge, Mass.: MIT Press, 1987); Michel Callon, "Techno-Economic Networks and Irreversibility," in *A Sociology of Monsters: Essays on Power, Technology and*

Domination, ed. John Law (London: Routledge, 1991), 132–61; Michel Callon, "Variety and Irreversibility in Networks of Technique Conception and Adoption," in *Technology and the Wealth of Nations: Dynamics of Constructed Advantage*, ed. Dominique Foray and Christopher Freeman (London: Pinter, 1993), 232–68; Michel Callon and Bruno Latour, "Unscrewing the Big Leviathan: How Actors Macro-Structure Reality and How Sociologists Help Them to Do So," in *Advances in Social Theory and Methodology: Toward an Integration of Micro- and Macro-Sociologies*, ed. Karen Knorr-Cetina and Aaron V. Cicouvel (London: Routledge, 1981), 277–303; Bruno Latour, *Science in Action: How to Follow Scientists and Engineers through Society* (Cambridge, Mass.: Harvard University Press, 1987); John Law, "Notes on the Theory of the Actor Network: Ordering, Strategy and Heterogeneity," *Systems Practice* 5, no. 4 (1992): 279–93; John Law, "After ANT: Complexity, Naming and Topology," in *Actor Network Theory and After*, ed. John Law and John Hassard (Oxford: Blackwell, 1999), 1–14.

28. Antonio Cordella, "Information Infrastructure: An Actor Network Perspective," *Journal of Actor Network Theory and Technological Innovation* 2, no. 1 (2010): 27–53; see also Eric Monteiro and Ole Hanseth, "Social Shaping of Information Infrastructure: On Being Specific about the Technology," in *Information Technology and Changes in Organizational Work*, ed. Wanda J. Orlikowski, Geoff Walsham, Matthew R. Jones, and Janice I. DeGross (London: Chapman and Hall, 1996), 325–43; Claudio U. Ciborra, ed., *From Control to Drift: The Dynamics of Corporate Information Infrastructures* (Oxford: Oxford University Press, 2001); Wanda J. Orlikowski and Susan V. Scott, "The Entanglement of Technology and Work in Organizations," LSE working paper series, 168, Information Systems and Innovation Group, London School of Economics and Political Science, London, 2008. Adopting similar assumptions, media scholar Emma Hemmingway used ANT to conduct an ethnographic study of television news production in the UK that highlighted a complex of sociotechnical relations that includes producers, anchors, reporters, and technicians, and extends from the in-studio media hub to the remote satellite truck. Emma Hemmingway, *Into the Newsroom: Exploring the Digital Production of Regional Television* (London: Routledge, 2008).

29. Karen Barad, *Meeting the Universe Halfway: Quantum Physics and the Entanglement of Matter and Meaning* (Durham, N.C.: Duke University Press, 2007).

30. Diana Coole and Samantha Frost, eds., *New Materialisms: Ontology, Agency, and Politics* (Durham, N.C.: Duke University Press, 2012), 9.

31. Jane Bennett, *Vibrant Matter: A Political Ecology of Things* (Durham, N.C.: Duke University Press, 2010), 25.

32. Bennett, *Vibrant Matter*, 37.

33. Donna Haraway, *Simians, Cyborgs, and Women: The Reinvention of Nature* (London: Routledge, 1990); Rosi Braidotti, *Metamorphoses: Towards a Materialist Theory of Becoming* (Cambridge: Polity, 2002); Rosi Braidotti, *The Posthuman* (Cambridge: Polity, 2013); Lucy Suchman, *Situated Actions: the Problem of Human-Machine Communication* (Cambridge: Cambridge University Press, 1987); Chela Sandoval, *Methodology of the Oppressed* (Minneapolis: University of Minnesota Press, 2000); Graham Harman,

Tool-Being: Heidegger and the Metaphysics of Objects (Chicago: Open Court, 2002); Ian Bogost, *Alien Phenomenology; or, What It's Like to Be a Thing* (Minneapolis: University of Minnesota Press, 2012).

34. Tarleton Gillespie, "The Politics of 'Platforms,'" *New Media Society* 12, no. 3 (2010): 347–64; and "Platform Politics," special issue of *Culture Machine* 14 (2013), available at, http://www.culturemachine.net/index.php/cm/issue/current (accessed September 18, 2014).

35. Colin McFarlane and Jonathan Rutherford, "Political Infrastructures: Governing and Experiencing the Fabric of the City," *International Journal of Urban and Regional Research* 32, no. 2 (June 2008): 371.

36. Lisa Parks, "Technostruggles and the Satellite Dish: A Populist Approach to Infrastructure," in *Cultural Technologies: The Shaping of Culture in Media and Society*, ed. Göran Bolan (London: Routledge, 2012), 64–84.

37. Stephen Graham and Simon Marvin, *Splintering Urbanism: Networking Infrastructures, Technological Mobilities and the Urban Condition* (London: Routledge, 2001), 33 and 382.

38. Ibid., 8; postscript.

39. Recent research has increasingly considered rural infrastructures. See Penelope Harvey, "Cementing Relations: The Materiality of Roads and Public Spaces in Provincial Peru," *Social Analysis* 54, no. 2 (2010): 28–46; Lisa Parks, "Where the Cable Ends: Television in Fringe Areas," in *Cable Visions: Television beyond Broadcasting*, ed. Sarah Banet-Weiser, Cynthia Chris, and Anthony Freitas (New York: New York University Press, 2007), 103–26; Fred Turner, "Burning Man at Google: A Cultural Infrastructure for New Media Production," *New Media and Society* 11, no. 1–2 (April 2009): 145–66; Darin Barney, "To Hear the Whistle Blow: Technology and Politics on the Battle River Branchline," *TOPIA: Canadian Journal of Cultural Studies* 25 (Spring 2011): 5–28.

40. Mark Poster, *The Mode of Information: Poststructuralism and Social Context* (Chicago: University of Chicago Press, 1990).

41. Manuel Castells, *The Informational City: Information Technology, Economic Restructuring, and the Urban-Regional Process* (Oxford: Blackwell, 1989), 146.

42. Christopher Blackwell and Gregory Crane, "Conclusion: Cyberinfrastructure, the Scaife Digital Library and Classics in a Digital Age," *Digital Humanities Quarterly* 3, no. 1 (2009); Geoffrey Rockwell, "As Transparent as Infrastructure: On the Research of Cyberinfrastructure in the Humanities," in *Online Humanities Scholarship: The Shape of Things to Come*, ed. Jerome McGann (Houston: Rice University Press, 2010); Patrik Svensson, "From Optical Fiber to Conceptual Cyberinfrastructure," *Digital Humanities Quarterly* 5, no. 1 (2011); Christine Borgman, *Scholarship in the Digital Age: Information, Infrastructure, and the Internet* (Cambridge, Mass.: MIT Press, 2007).

43. See for example, Pippa Norris, *Digital Divide: Civic Engagement, Information Poverty, and the Internet Worldwide* (Cambridge: Cambridge University Press, 2001); Philip N. Howard, "Testing the Leap-Frog Hypothesis: The Impact of Existing Infrastructure and Telecommunications Policy on the Global Digital Divide," *Information, Communi-*

cation and Society 10, no. 2 (April 2007): 133–57; Anna Everett, *Digital Diaspora: A Race for Cyberspace* (Albany, N.Y.: SUNY Press, 2009).

44. Dennis Rodgers and Bruce O'Neill, "Infrastructural Violence: Introduction to the Special Issue," *Ethnography* 13 (2012): 402.

45. Ibid., 404.

46. Nigel Thrift, "Remembering the Technological Unconscious by Foregrounding Knowledges of Position," *Environment and Planning D: Society and Space* 22 (2004): 177.

47. Carolyn Marvin, *When Old Technologies Were New: Thinking about Communications in the Late Nineteenth Century* (New York: Oxford University Press, 1988); Greg Downey, *Telegraph Messenger Boys: Labor, Technology, and Geography, 1850–1950* (New York: Routledge, 2002); Claude S. Fischer, *America Calling: A Social History of the Telephone to 1940* (Berkeley: University of California Press, 1994); Brian Larkin, *Signal and Noise: Media, Infrastructure, and Urban Culture in Nigeria* (Durham, N.C.: Duke University Press, 2008).

48. Stephen Graham, ed., *Disrupted Cities: When Infrastructure Fails* (New York: Routledge, 2010).

49. Stephen Graham, "When Infrastructure Fails," in *Disrupted Cities*, ed. Stephen Graham (New York: Routledge, 2010), 3.

50. Jackson, "Rethinking Repair," in *Media Technologies: Essays on Communication, Materiality and Society*, ed. Tarleton Gillespie, Pablo Boczkowski, and Kirsten Foot (Cambridge, Mass.: MIT Press, 2014), 222.

51. Lisa Parks, "Media Fixes: Thoughts on Repair Cultures," *Flow* (December 2013), available at http://flowtv.org/2013/12/media-fixes-thoughts-on-repair-cultures (accessed September 18, 2104); Daniela Rosner, "Making Citizens, Reassembling Devices: On Gender and the Development of Contemporary Public Sites of Repair in Northern California," *Public Culture* 26, no. 1 (2014): 51–77; Daniela Rosner and Fred Turner, "Theaters of Alternative Industry: Hobbyist Repair Collectives and the Legacy of the 1960s American Counterculture," in *Design Thinking Research: Building Innovators*, eds. Hasso Plattner, Christoph Meinel, and Larry Leifer (Cham, Switzerland: Springer, 2015), 59–69.

52. Richard Maxwell and Toby Miller, *Greening the Media* (Oxford: Oxford University Press, 2012), 29.

53. Nadia Bozak, *The Cinematic Footprint: Lights, Camera, Natural Resources* (New Brunswick, N.J.: Rutgers University Press, 2012), 3.

54. Jussi Parikka, ed., *Medianatures: The Materiality of Information Technology and Electronic Waste* (London: Open Humanities Press, 2011); Richard Maxwell, Jon Raundalen, and Nina Lager Vestberg, eds., *Media and the Ecological Crisis*, Routledge Research in Cultural and Media Studies (New York: Routledge, 2014).

55. Sarah Kember and Joanna Zylinska, *Life after New Media: Mediation as a Vital Process* (Cambridge, Mass.: MIT Press, 2012), 23.

56. Ashley Carse, "Nature as Infrastructure: Making and Managing the Panama Canal Watershed," *Social Studies of Science* 42 (2012): 539–63.

57. Patricia Clough, *The Affective Turn: Theorizing the Social* (Durham, N.C.: Duke University Press, 2007).

58. Greg Seigworth and Melissa Gregg, *The Affect Theory Reader* (Durham, N.C.: Duke University Press, 2010), 2.

59. Darin Barney, "To Hear the Whistle Blow," 7.

60. Ibid, 8.

61. Ibid, 7.

62. Bowker, et al., "Toward Information Infrastructure Studies," 113.

63. Brian Larkin, "The Politics and Poetics of Infrastructure," *Annual Review of Anthropology* 42 (2013): 339.

64. Star and Ruhleder, "Ecology of Infrastructure."

PART I

Compression, Storage, Distribution

CHAPTER 1

Compression

A Loose History

JONATHAN STERNE

The use of the word *compression* to describe a communication technology process comes rather late in its history. According to the *Oxford English Dictionary*, the term *compression* is at least six hundred years old. Its use to describe the "condensation of thought and language" dates to the eighteenth century. The term was first applied to machinery—steam engines—in the mid-nineteenth century. Compression as a description of representation thus predates its use to describe a technical operation by about one hundred years.[1]

Today, compression in communication engineering refers to one of two things: data compression or dynamic range compression. People encounter data compression every day in the form of zipped files, mp3s, jpegs, online videos, and mobile-phone voice algorithms. All of these technologies save precious bandwidth by eliminating categories of data that engineers have decided are redundant and therefore unnecessary to store or transmit. Dynamic range compression refers to reducing the distance between the loudest and quietest parts of an audio signal. It is useful because a signal with less variance can have a higher overall average volume.

Most writers outside the engineering world, and especially most humanities scholars, still understand compression as something that happens after the fact, as supplemental to communication and its purposes, to perception, to interaction, and to the experiences attending them. In the wake of poststructuralism, few humanities writers would argue for verisimilitude[2] as a guiding norm for representation—whether technological or otherwise. Yet too often we still write our media theories and histories as if the primary aesthetic criteria for technical media are verisimilar. In such work media representations are judged not only in terms of their realism but also in terms of their self-sufficiency; perceptual and definitional abundance; and immersive characteristics. This is the story of communication as being about the anxiety over the loss of meaning through a succession of technical forms. The assumption here is that progress in technology comes through its ability to produce verisimilitude.

For example, we can hear this set of assumptions as the warrant behind an implicit criticism of the sound of contemporary music in a *New York Times* story about vinyl records:

> The last decade has brought an explosion in dazzling technological advances—including enhancements in surround sound, high definition television and 3-D—that have transformed the fan's experience. There are improvements in the quality of media everywhere—except in music. In many ways, the quality of what people hear—how well the playback reflects the original sound—has taken a step back. To many expert ears, compressed music files produce a crackly, tinnier and thinner sound than music on CDs and certainly on vinyl. And to compete with other songs, tracks are engineered to be much louder as well.[3]

But it is not only journalists who argue this way, and we should be grateful, since academics are more likely to render their assumptions explicitly. Writing about optical devices like the telescope and microscope in the nineteenth century, Anne Friedberg argued that their "entertainment function . . . relied not only on the verisimilitude of the images seen and the recording capabilities of mediated vision, but also on the illusion of verisimilitude, the very *virtuality* of the experience produced."[4] The comment is interesting both because Friedberg explains her own logic and because if presented with the proposition in the abstract—*verisimilitude* is the basis of virtuality and entertainment—as an author influenced by poststructuralist thought, she would not, I suspect, accept the proposition.

In writing about technical media, verisimilitude is often tied to signal definition. Definition is the amount of signal that can fit in a given transmission or be stored in a file. It is the available bandwidth or storage capacity of a medium in terms of how much of its content can be presented to an end user at any given moment. It measures the density of materials available to perception; *available* is the key term here, because signal definition guarantees neither robustness of perception nor intensity of experience for listeners or viewers. The number of pixels on your screen is a measure of definition; high-definition television is measured in pixel density. The number of bits-per-second transmitted as your digital audio file plays back is also a measure of definition. It is a traditional line of marketing rhetoric to assert that increased signal definition leads to increased realism in the representations of a given medium or format, and that the increased realism will lead to greater intensity of experience and deeper meanings for audiences. But this is not actually the case. As Michel Chion writes,

> Current practice dictates that a sound recording should have more treble than would be heard in the real situation (for example when it's the voice of a person at some distance with back turned). No one complains of nonfidelity from too much definition! This proves that it's definition that counts for sound, and its hyperreal effect, which has little to do with the experience of direct audition.[5]

Analogous arguments can be made for images and video: definition is not verisimilitude, definition is not realism, and realism is not reality, but these terms are still often confused.

There is an aesthetic tradition that goes in the opposite direction. If we want to understand compression as a cultural phenomenon, as something other than a perversion or diminishment of more primary, higher-definition sensory experience, it would be wise to begin with how end users experience it. Aesthetics matter here for several reasons: so much of the writing that orbits around verisimilitude makes aesthetic arguments, and so if I want to pose a viable alternative, I need to at least gesture in that direction. We also tend to think of the storage and transmission dimensions of media as anaesthetic phenomena, mere engineering matters—more *telecommunications* than *communication*. But in fact they are absolutely central to culture, experience, and action at a distance. They help shape the texture of mediatic experience.

Consider this account of telegraphic conversation from a mid-nineteenth-century piece of short fiction, "Kate: An Electro-Mechanical Romance":

> Mary replied instantly, and at once the two girl friends were in close conversation with one hundred miles of land and water between them. The conversation was by sound in a series of long and short notes—nervous and staccato for the bright one in the little station; smooth, legato and placid for the city girl. . . .
>
> [T]he two friends, one in her deserted and lonely station in the far country, and the other in the fifth story of a city block, held close converse . . . for an hour or more, and then they bid each other good night, and the wires were at rest for a time.[6]

Here, the basis of intensity and intimacy is precisely a lack of definition. Whole modes of being are condensed into the rhythms of telegraph signals, which in turn index the subtle and quick movements of operators' hands. We could attribute this description to a standard nineteenth-century literary conceit were it not so common elsewhere. In *When Old Technologies Were New* Carolyn Marvin tells stories of telegraphic and telephonic weddings and deceptions, feats of long-distance intimacy and intensity, whether in shared passion or cruelty.[7]

Frantz Fanon's treatment of radio in *A Dying Colonialism* follows a similar pattern. In his chapter, "This Is the Voice of Algeria," he writes, "The whole nation would snatch fragments of sentences in the course of a broadcast and attach to them decisive meaning. Imperfectly heard, obscured by an incessant jamming, forced to change wave lengths two or three times in the course of a broadcast, the *Voice of Fighting Algeria* could hardly ever be heard from beginning to end. It was a choppy, broken voice."[8] Lucas Hilderbrand coins the phrase "bootleg aesthetics" to describe the grainy images of analog video that gave rise to fair-use law and presaged many file-sharing practices in the United States. He, too, finds these videos all the more affectively powerful because of their low definition. Blurred images and distorted sounds could serve as material traces of a video's illicit circulation, adding a potential thrill or at least a call to identify with countercultures of circulation.[9]

Limited definition can produce particularly intense modes of experience. *Intense experience shaped by limited definition.* This is an old point from Marshall McLuhan. In his classic essay "Media Hot and Cold," McLuhan discusses definition as an affective problem, rather than a reality problem. "A hot medium is one that extends a single sense in 'high definition.' High definition is the state of being well filled with data." Cool media are *low definition* "because so little is given and so much has to be filled in." With the television image, he writes, the eye must "act as hand in filling in and completing the image."[10] Derived as it was from the everyday experience of watching black-and-white images flicker on cathode ray tubes and hearing sounds emanate from tiny monaural speakers with cheap transistor amplifiers, McLuhan's description of television as cool might well have felt ontological to the end user of 1964. Today, the range of television experiences available to the average person—from mobile-phone screens to HD—reveals television's coolness as a specifically infrastructural, industrial, and cultural condition. Coolness was an aesthetic that tuned perception to the limits of transmission infrastructure, and tuned transmission to the then-understood limits of perception. Using McLuhan's terminology, to say all media follow an historical trajectory toward high definition, to write media history in terms of a general history of verisimilitude, is to say that all are on a historical path toward hotness. But this is clearly not the case, either in his time or ours. Whether in its audio or data varieties, compression accommodates signals to infrastructures. But it also transforms infrastructures by enabling them to carry different kinds of signals.

More generally, we can define compression this way: compression is the process that renders a mode of representation adequate to its infrastructures. But compression also renders the infrastructures adequate to representation.[11] There are thus at least two long-term tendencies in media history, and at least two grand narratives of media history we need to consider (and almost certainly more).[12] The dominant paradigm is the general history of verisimilitude,[13] where progress in technological history is therefore imagined as progress in terms of greater and greater definition. Too often, humanist critics of media forms echo advertising copy, as sound and images are unmoored from previously fixed stations governed by immediate experience, or reduced in definition or sophistication by virtue of their technological transmission (but with the promise of better definition—and more realism—in the next generation of technology).[14] Humanists still write like travelers who want to anticipate every possible social or climatological contingency at their destination. The clothes preexist the trip—they take up less space in the suitcase if we roll them up and squeeze out the extra air to allow for a few extra garments to make the journey with us. In this line of thinking, compression squeezes the extra air out of sound recordings, phone calls, or videos that would otherwise take up more space as they reach out toward a horizon of fulsomeness. This chapter sketches·out *a general history of compression* as another path through media history, and gives a rough outline of its contours. A general history of compression considers communication as based in a relational reality (often, though not necessarily, networked) and presupposes large-scale, collective activity in its positivity. Following Gilbert Simondon, there is a particularly useful insight from a history of compression for thinking about communication in general. Starting from compression, communication has a "network reality."[15] This is to say that it is not a binary relationship between sender and receiver mediated by a medium but rather an ensemble of relations that only produce the moments of transmission and reception after the fact. For Simondon this is a kind of circular causality, or at least a relational causality, where a relation must exist to produce the things on which it has effects. As he argues in *On the Mode of Existence of Technical Objects:*

> Elements that materially are to constitute the technical object, and that are independent one of the other, lacking an associated milieu that precedes the constitution of the technical object, must be organised in relation to one another by means of *circular causality* which will exist once the object is constituted. What is involved here, then, is a conditioning of the present by the future, or by what up to now does not exist.[16]

Simondon's prime example of this process at work is the Guimbal dam in the Philippines, where the water moved by the turbine cools the turbine, enabling it to work at the properly regulated temperature to move water.[17] For media scholars, this means that media are not like suitcases; and images, sounds, and moving pictures are not like clothes. They have no existence apart from their containers and from their movements—or the possibility thereof. Compression makes infrastructures more valuable, capable of carrying or holding materials they otherwise would or could not, even as compression also transforms those materials to make them available to the infrastructure. Having taken a short detour through theory, let us now take a second detour through the history of technology, to consider how compression gets named as a problem in the twentieth century.

* * *

In the field of communication technology, compression was first applied to audio. The *Oxford English Dictionary* gives 1938 as the earliest known use of the term *compression* as it is applied to audio. The December 1937 issue of the magazine *Communications* offers further insight into the term's use at the time in the United States. "Compressors" reduced the distance between the loudest and quietest parts of audio signals. Engineers started to call them "limiters" because of how they worked. Today this technique is called "dynamic range compression," and particular types of compression are called "limiting." The *Communications* article discusses the Western Electric 110-A amplifier and the products of two competitors, all of which had come onto the market in the previous year. In less than a year, "over half the radio stations in the country" had purchased one.[18] To understand why radio stations rushed to buy these devices, we have to understand a bit about loudness and radio.

Compression solved a problem created by the Federal Communication Commission's method of regulating radio stations after 1927. Each station was allotted a certain channel and a certain maximum broadcasting power. Exceeding this allotted power, even for a moment, was called "overmodulation"—it was against regulation and it was considered rude. As explained by John P. Taylor, author of the *Communications* article, overmodulation introduced various kinds of signal distortion into the sound, but it also interfered with adjacent channels.[19] It was the equivalent of shouting down your neighbor.

At the same time, the upper limit placed on broadcast power introduced certain aesthetic problems for radio broadcasters. Roughly speaking, broadcast power in wattage was a measure of how loud a station could be relative to other stations on the dial. Stations wanted to be as loud as possible for as much of the time as possible. This was part of a logic of capitalist competition, where it was expected that the consumer flipping through the dial would

Figure 1.1. Western Electric 110A amplifier

gravitate toward the louder signal. Although watts are not a very good measure of loudness, peak wattage represented peak loudness, and all other levels were relative. The range of possible loudnesses was called dynamic range. So if a station's average levels were quite a distance below its peak levels, most of the time it would not be using all of its allotted wattage, and that station would be effectively too quiet compared with its competitors, placing it at a commercial disadvantage.

Explaining limiters, the 1937 *Communications* article sounds a lot like Bruno Latour discussing delegation in his famous door-closer essay, where the automatic door-closer replaces someone hired to keep a door closed:[20]

> At first glance, the operations performed by a limiting amplifier seem much like those of the studio control operator in riding gain [by manually turning a volume knob]. The only actual similarity, however, is in that both entail compression of the volume range. The function of the control operator is to manually adjust the gain in accordance with the average level of the program. He can raise a low passage, or reduce a loud one; but he cannot, as a rule, act quickly enough to cut

down loud peaks of short duration. If he is adept in following the score, or has had the benefit of many rehearsals, he may indeed anticipate some of these. But most operating is not accomplished under such conditions, and even when it is, there will remain occasional peaks not suppressed as desired. . . . [Therefore] since these peaks may represent levels several times the average level, such operation necessarily means very ineffective use of the available power [the wattage allotted by the FCC]. It is this situation which the limiting amplifier is intended to improve. The method is to provide a gain reduction system which, coming into play at high audio levels, automatically reduces the gain of the system and thereby keeps peak levels within a predetermined limit. Properly used, this allows the average modulation to be stepped up to something like half again as much—say from 30 percent to 45 percent. Output at the receiver is, of course, proportionately increased.[21]

As a form of audio limiting, compression solved a problem of infrastructure for radio engineers. It transformed the signal, but crucially, it also made the infrastructure more convivial to the broadcast signal. The radio engineer who manually adjusted volume listened for parts of a program—scenes of a play, musical movements—and operated in terms of seconds and minutes. The compressor detected volume peaks on the order of tenths and hundredths of seconds. Rather than sections of works, it addressed the consonants in words, the crashing of cymbals, or any other transient characteristics of sounds. It works on a completely different timescale, and this smaller timescale produces a much more even signal over time. Since the FCC regulated radio in terms of loudness, and since stations competed with one another in terms of loudness, a device that could increase the average loudness of a broadcast was immensely valuable. It meant they could use more of their allotted bandwidth, and use it more effectively.

As the *Communications* article notes, commercial compression devices came late to radio. Military applications had existed since the mid-1920s, and custom devices similar to the 110A had been in use in motion-picture sound for several years, because of the extremely limited dynamic range of optical recording onto film as an audio format. Over the next few decades, compressors would find other uses as well. Karin Bijsterveld notes that they were used in factories to increase average volume of piped-in Muzak and to overcome ambient noise levels by brute force. Professional recording studios also took them up first for practical reasons, but they quickly became a prized part of the sound of midcentury and later popular music.[22] Today, the digital modeling of analog compression is one of the holy grails of digital audio processing, a story I will tell elsewhere.

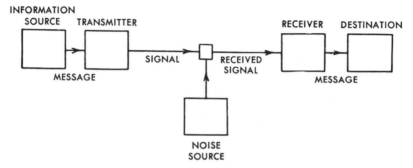

Figure 1.2. Shannon's schematic for his Mathematical Theory of Communication: it was presented as pertinent to all forms of communication, a general theory.

This capsule history of dynamic range compression outlines a pattern. While not universal, it offers a heuristic and a set of questions for thinking through compression as a recursive dimension of communication history. Of course, like any abstract relation, it is only available in the concrete, and any concrete manifestation will play out differently.[23] But remember the project here: if the general history of verisimilitude conceives of media in terms of their possibility to transcend the problems of representation and achieve a full identity between original and copy (and the inevitable failure of that transcendence), the general history of compression instead asks how media manage and enact relations shaped by one or more conditions of finitude:

1. Technological communication is always engineered, and compression is the mode through which engineers negotiate the limits and affordances of an infrastructure. If this is so, how are limits—technical, perceptual, juridical, cultural—negotiated in a given assemblage of practices, technologies, institutions, and representations?

2. We normally think of compression as the conditions of an infrastructure operating on a signal, but it also marks a signal operating on an infrastructure. Compression effects both a transformation of the signal and the meaning and utility of the infrastructure it inhabits. If this is so, how do compression and content make a given infrastructure possible?

3. Engineers assess this operation in the technical domain, but users (broadly defined) experience compression in the aesthetic domain—less as part of the character of an infrastructure (to the extent that people notice the infrastructure at all[24]) than the character of the sounds and images it transmits. If this is so, how

do we understand aesthetics and experience beginning from an assumption of finitude, rather than comparing it with an imaginary standard of transcendence?

Compression history posits communication and representation as relational phenomena first and foremost. It is also not particularly idealistic about the content or situation of communication. Borrowing from Simondon again, to consider representation from the standpoint of compression, rather than from verisimilitude, means to consider it "in its entelechy, and not in its inactivity or static state."[25]

Let us now pose these three historical questions to data compression, the form more familiar to most people today. Usually, data compression is understood as a completely separate process that simply shares a name with audio compression. Wikipedia, for instance, considers them to be separate processes. But they are historically and technically related, and the evidence is in one of the founding documents of data compression: Claude Shannon's classic 1948 *Mathematical Theory of Communication.* Shannon refers to compression three times, and, over the course of the book, he directly links the different senses of compression considered thus far: condensation of thought and language, audio compression, and data compression.

Shannon's first use of the term *compression* is to describe the transmitter part of his famous "communication system" diagram. Speech must be "compressed," he says, before it is transmitted by a transmitter. It is also worth noting that Shannon's transmitter is defined infrastructurally as that which "operates on the message in some way to produce a signal suitable for transmission over the channel."[26] This usage is essentially the same as the radio engineers. Compression here is about suitability for movement through an infrastructure.

But later, he uses compression in the sense now used for data compression, that is, removing redundant data from source material:

> The redundancy of ordinary English, not considering statistical structure over greater distances than about eight letters, is roughly 50%. This means that when we write English half of what we write is determined by the structure of the language and half is chosen freely. The figure 50% was found by several independent methods which all gave results in this neighborhood. [Shannon goes on to explain how he arrived at this figure.]
>
> Two extremes of redundancy in English prose are represented by Basic English and by James Joyce's book *Finnegans Wake.* The Basic English vocabulary is limited to 850 words and the redundancy is very high. This is reflected in the expansion that occurs when a passage is translated into Basic English. Joyce on the other hand enlarges the vocabulary and is alleged to achieve a *compression* of semantic content.[27]

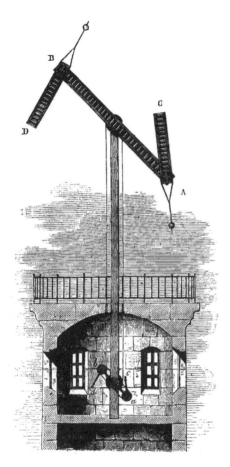

Figure 1.3. Chappe telegraph. A series of levers inside the tower allowed the operator to move the arms into different positions, which could be looked up in a codebook. Messages could thus be rapidly relayed across long distances, using lines of visibility.

Here it is, all in a single, neat bundle: Shannon's example for mathematical compression *is* the 1400s meaning of "condensation of thought and language."

What we would today call data compression appears again in his discussion of discrete noiseless systems (which are a mathematical idealization of telegraphic modes of communication) where he explains the math behind removing redundant elements of a message, be they letters in the alphabet or numbers in the calculation of p_i.[28] Although Shannon's discussion sounds highly technical, it is based on a very old idea. The omission of vowels in Hebrew writing or the modes of abbreviation in a medieval music manuscript all assume a knowledgeable community of interpretation. In Shannon's terms, part of the transmission is redundant. Today, data compression and dynamic range compression are understood as two entirely different things. But it is clear from this history that, in fact, they are intimately interrelated, along with the very old notion of condensation of thought, language, and experience.

* * *

Despite the fact that the process was only named in the twentieth century, the idea of compression names a much longer-term phenomenon in communication history. For instance, of long-distance communication in antiquity, Harold Innis writes, "Cheap copies of works suited to Christians as a persecuted sect were probably written on papyrus in the form of a codex in the second century AD. The codex form carried more material and was more convenient. Use of the roll would restrict content to the single Gospels of the Acts."[29] Compared with the scroll, the codex was a form of compression. Space is expensive, and weight put practical limits on transmission in the physical world. In the electronic world, often the most valuable and expensive component is not the infrastructure itself but bandwidth—as every mobile-data subscriber knows. Communication systems are organized around their capacities to transmit or store (or both), so it is no surprise that capacity is where the strongest economic pressures lie.

The long history of telegraphy shows how content and infrastructure exist in a relation of circular causality. Telegraphs only work if they have codes. Though the earliest long-distance media were not telegraphs themselves, they

Figure 1.4. Cooke and Wheatstone's telegraph. Combinations of needles pointed to letters.

can be characterized retrospectively in telegraphic terms. Ancient military communication technologies like flags, trumpets, drums, and beacon signals that used fire and smoke all had very limited repertoires of possible signals and messages.[30] Later, optical telegraphs took advantage of innovations like telescopes (thus increasing distance between stations) and automation. But the most important telegraphic innovations were more elaborate codes. The most famous one, Claude Chappe's optical telegraph in France (originally called the *tachygraphe*) made use of telescopes and multiple moving arms, but they were rendered functional by their elaborate codebooks. Capable of 196 possible positions (ninety-eight for messages and ninety-eight for regulating the line), the telegraph *could* have indicated individual letters easily, as a subsequent British optical telegraph did.[31]

To make transmission more efficient, Chappe's cousin devised a codebook consisting of 9,999 "words, phrases and expressions each represented by a number," though eventually the total number of entries was reduced to speed look-up.[32] In practice, this meant attaching a number to each position of the arms, so that for instance "a message of 2, 15, 88 meant the 88th word on page 15 of book 2."[33] Terminal operators could then decode the message using their codebooks.

Optical telegraphers could thus convey relatively long messages with just a little content moving through the system. Countless media historians have cited electric telegraphy as the first modern communication medium and have attributed to it all sorts of significant innovations and effects.[34] Here we can learn a bit by minding our dates. The electric telegraph appeared as a possibility in 1753, when Charles Morrison anonymously published an article as "C.M." in *Scots Magazine*. But Morrison had not worked out an effective code, and other early electrical telegraphs used letters of the alphabet.[35] The Chappe mechanical system, along with several others, was developed in the 1790s as a more robust system precisely because of its code. It took Morse's system of dots and dashes, and Cooke and Wheatstone's system of needles and letters, to make electric telegraphs economically viable.[36]

Crucially, both Cooke and Wheatstone's and Morse's codes had elements of compression in them. The original five-needle Cooke and Wheatstone telegraph left out the letters C, J, Q, U, X, and Z. Morse's code originally contained only numbers, which corresponded to words that could be looked up in a codebook. Alfred Vail improved his system by adding letters. But he also organized the system through a compression scheme. After visiting the local newspaper print shop in Morristown and counting the frequency of letters in their type cases, he gave the most frequently used letters the simplest representations in his code

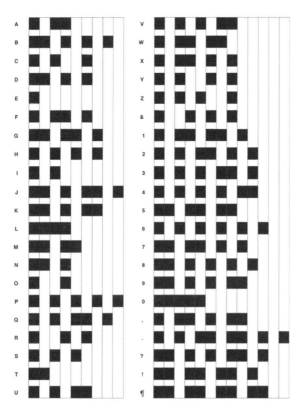

Figure 1.5. American Morse Code. Morse's telegraph produced dots or dashes in a roll of tape, depending on how long a circuit was connected. Later, telegraphers discovered that they could listen to the device for even more rapid transmission.

(for instance, *E* received a single dot).[37] When Morse operators started listening to their receivers rather than reading them, that also sped up the process.[38] And when operators reintroduced Morse's appropriation of the codebook on top of his code, a whole second layer of compression occurred.

In *How We Think*, Katherine Hayles writes, "Constructed under the bywords 'economy,' 'secrecy,' and 'simplicity,' [electric] telegraph code books matched phrases and words with code letters or numbers. The idea was to use a single code word instead of an entire phrase, thus saving money by serving as an information compression technology."[39] The electric telegraph codebook's status here is ambiguous. Hayles considers it a compressive supplement to a previously existing system. But if we think of the long history of optical telegraphy, it is clear that a codebook, or a code, is necessary for the telegraph to work at all. Even in Morse's case, the codebook was historically prior to the alphabetic code. As Hayles writes, every institution had its own codebook on top of Morse code. She notes a circular causality between content and infrastructure. Like limiting amplifiers in 1937, codebooks rapidly proliferated across users. Perhaps both Morse and Cooke and Wheatstone's alphabetic telegraphs, though already compressed, were themselves historical exceptions. If we were to choose a random point in its various histories, a "normal" telegraph would more likely work according to a codebook or its equivalent.

There is a particular logic to a codebook worth considering. We can hear its echoes in Shannon's appeal to the 850 words of Basic English, and his interest in 50 percent of all letters being predictable. Here is Hayles's clever description of a codebook in its actual use:

> Imagine you are sitting in your office preparing to write a telegram. Your employer, ever conscious of costs, has insisted that all office telegrams must be encoded. . . . If the phrasing deviates slightly from what you might have intended, you are willing to accept it as "good enough," because it would take time and money to write out in plain text a hypothetical alternative. To make the example more specific, suppose you work for a banking firm and have heard a rumor that a client asking for a loan is in financial trouble. Using the *Direct Service Guide Book and Telegraphic Cipher* (1939), you find under the keyword "Difficulties Rumored," the code word BUSYM, standing for "We have information here that this concern is in financial difficulties. An immediate investigation should be made. Send us the results." You had not intended to ask for an investigation. . . . But seeing the phrase, you think perhaps it is not a bad idea to press for further information. You therefore write BUSYM and send off your telegram, confident it expresses your thoughts. This fictional scenario suggests that the code books, by using certain phrases and not others, not only disciplined language use but also subtly guided it along paths the compilers judged efficacious.[40]

The "good enough" dimension of communication is of interest here. In the codebook scenario, communication is meant to be efficacious rather than verisimilar. If you pause to consider your average day, you will find that this is in fact a more common state of communication in general. Thus, Hayles argues, codebooks "were [on the one hand] used in straightforward business practices to save money. On the other hand, through their information compression techniques, their separation of natural-language phrases from code words, and the increasingly algorithmic nature of code construction," they precede the modern information economy.[41] Considered as part of compression history, the "straightforward business" practices she mentions are built around multiple materialities of information, from bandwidth to the modes of arranging and experiencing that information. Telegraphic code physically occupies bandwidth and storage, it takes one form or another, and it costs money, tying it to capital in its basic operations.

One of the key lessons of compression history is to orient humanistic inquiry toward telecommunications systems as themselves substantive concerns. Hayles moves in this direction when she examines the codebooks, but it is possible to extend further her argument. She later writes, "What if language, instead of sliding along a chain of signifiers, were able to create a feedback loop of continuous reciprocal causality such that the mark and concept co-constituted each other?"[42] In the figure of the telegraph code and especially in codebooks that serve a function we would today call aggregation, the mark, the concept, and the infrastructure are already co-constituted.[43]

We could say the same of images and sounds. Halftone printing epitomizes some of the advantages of compression and formed the material basis of an emerging nineteenth-century visual culture. Appearing first in magazines and later in newspapers, halftones allowed for photographic reproductions to proliferate in late-nineteenth-century print media, and after 1910 a fast, efficient block-printing technique built around the halftone process supplanted older practices. As halftones became cheaper and more plentiful, images could compete with words for priority on the printed page. Composed of hundreds or thousands of tiny dots that vary in size, shape, or layout, halftones rely on readers' eyes blurring the dots into a continuous image at a distance from the page. A multiple-halftone process facilitated color prints of photographs, and halftones share a technical logic with facsimile transmission and pixel-based display technologies. Besides suggesting a common material basis for paper- and screen-based media, halftone history illustrates the way that even something as light and portable as paper can be made more dense, more available to new kinds of content through compression techniques. "Every commentator on the magazine revolution mentions the halftone press, whose development made it possible to print reproductions of photographs quite cheaply, and on type-compatible paper," writes Richard Ohmann.[44]

Early attempts to introduce efficiencies into the telephone system followed a similar logic to the telegraph, but they also brought auditory perception more fully into the fold. Modern speech-and-hearing science emerged out of the desire to understand the minimum amount of signal that could be transmitted over a phone line and still be understood as intelligible speech. As Mara Mills has pointed out, Homer Dudley's work on the vocoder, which preceded Shannon's mathematical theory, already operated on the principle that a large portion of language—and specifically speech—was redundant to communication. I use the term *perceptual technics* to name the application of perceptual research for the purposes of economizing signals. For Bell Labs, the point of understanding human speech and hearing was to determine minimal thresholds of intelligibility. If the bandwidth in the phone system exceeded the amount needed to reproduce speech, then the phone company could repurpose that bandwidth to carry other calls. This is more or less what happened: based on a combination of perceptual technics and new filtering devices, Bell quadrupled the carrying capacity of the existing infrastructure. Economy and efficiency shape both the medium and perception, but the endpoint is Virginia Heffernan's nostalgia for her friends' telephone voices.[45]

Analog color television in the United States also follows the compression story. Faced with the challenge of fitting a wider-bandwidth color signal into the existing infrastructure for U.S. television, engineers at RCA and the National Television Systems Committee compressed the color TV signal by removing "redundant" content in the manner described by Shannon, and followed by the phone system. The existing psychophysics told them that the eye has more acuity for green than for red and blue, and so the color signal reproduced most of the green, some of the red, and a little of the blue in order to fit into the existing available bandwidth.[46]

This notion of compression is the one that drives the development of now-ubiquitous encoding schemes for digital audio and video. Almost every image and sound that comes through a digital format has encountered some kind of compression. Many digital encoders use an algorithm developed by David Huffman (a student of Shannon's collaborator Robert Fano) to eliminate redundant data and then use a mathematical model of perception based on the principle that much of the content rendered by a full-definition audio system is not actually perceived by most viewers or listeners in most situations. Like telegraphs, radios, and analog television sets, digital audio and video treat communication as a "network reality," except here Simondon's network metaphor starts to get strained, since the listener exceeds any narrowly defined network. We might simply call the reality relational. For instance, a single set of standards like those set by MPEG, the Moving Picture Experts' Group, facilitated the circulation of video and audio recordings on the Internet, but they also facilitated the development of new technologies of storage and transmission, like the video compact disc, satellite radio, and the DVD. Once again, it is not just communication adjusting to infrastructures, but infrastructures modified by phenomena of compression. We could say the same of technologies of storage and transmission in general.

One could easily fold this story back into the history of verisimilitude. I could end by claiming we are getting ripped off or cheated out of definition, whether we are talking about data density, definition, or dynamic range; that "the man" does not want us to have all our definition. But this would be wrongheaded. One could imagine a host of "good enough" effects: the phonetic alphabet—alpha tango foxtrot—but also mobile-phone algorithms and YouTube. Again, entelechy is the key. If we consider these technologies in their use, rather than as diminished instantiations of an imaginary ideal, another sense emerges. What if we followed the old tradition of cultural criticism in writers like Raymond Williams, reading aesthetics as expressions and mediations of cultural conditions?[47]

Writing about digital audio reproduction in 1987, John Mowitt noted that as digital channels produced less and less audible distortion, musicians sought out more and more means of generating distortion for aesthetic purposes. That was years before the current analog revival and fashion for digitally modeling analog devices. The explosion of work on noise has in part tried to push back against this tendency. For instance, Peter Krapp begins his *Noise Channels* by arguing that a large swath of digital culture embraces "the reserves that reside in noise, error and glitch." Following Michel Serres, writers like Greg Hainge have made a virtue of noise. But some of what is considered noise is simply a compression artifact. There is probably also a story to be told about what David Harvey famously called time-space compression, which is intimately and phenomenally linked to the expansion of massive transportation and communication infrastructures.[48]

The omnipresence of distortion in popular music as well as in sonic art is about the overloading and saturation of channels. In Alvin Lucier's *I am Sitting in a Room* the space becomes a musical instrument. We could say the same of the distortion inherent in a Tito Puente recording, in the dulled attack of a Rhodes electric piano, or in distorted voices on a bullhorn or feeding-back guitars. Even much so-called glitch music is created simply by filling up a channel until it hits its limit, until the software says "no more." In the textures of their compression practices, artists and musicians may have something on media theorists. For in practice, they consider a limit as something that produces representation, rather than interfering with it. Like the old existentialist line about death, understanding communication might start not with plenitude. It might begin with finitude.

Notes

Thanks to the audiences at NYU, Harvard, and Northwestern, and to the editors and Dylan Mulvin. Thanks also to the Social Sciences and Humanities Research Council of Canada, whose funds supported some of this work.

1. *Oxford English Dictionary*, s.v., "compression."

2. Although the term is usually linked to life-likeness, etymologically the term is closer to "truthiness," to use a wonderful term from the noted philosopher Stephen Colbert.

3. Joseph Plambeck, "In Mobile Age, Sound Quality Steps Back," *New York Times*, May 9, 2010.

4. Anne Friedberg, *The Virtual Window: From Alberti to Microsoft* (Cambridge, Mass.: MIT Press, 2006), 63, emphasis in original.

5. Michel Chion, *Audio-Vision: Sound on Screen*, trans. Claudia Gorbman (New York: Columbia University Press, 1994), 98–99.

6. Charles Barnard, "Kate: An Electro-Mechanical Romance," in *Lightning Flashes and Electric Dashes: A Volume of Choice Telegraphic Literature, Humor, Fun, Wit, and Wisdom, Contributed to by All the Principal Writers in the Ranks of Telegraphic Literature as Well as Several Well-Known Outsiders, with Numerous Wood-Cut Illustrations*, ed. W. J. Johnston (New York: Johnston, 1877), 56–57.

7. Carolyn Marvin, *When Old Technologies Were New: Thinking about Electrical Communication in the Nineteenth Century* (New York: Oxford University Press, 1988).

8. Frantz Fanon, *A Dying Colonialism*, trans. Haakon Chevalier (New York: Monthly Review, 1965), 86

9. Lucas Hilderbrand, *Inherent Vice: Bootleg Histories of Videotape and Copyright* (Durham, N.C.: Duke University Press, 2009), 178. This point echoes Brian Larkin's reading of video in Brian Larkin, *Signal and Noise: Media, Infrastructure, and Urban Culture in Nigeria* (Durham, N.C.: Duke University Press, 2008). Nostalgia for landline phones follows a similar pattern. See, for example, Virginia Heffernan, "Funeral for a Friend," *New York Times*, October 31, 2010.

10. Marshall McLuhan, *Understanding Media: The Extensions of Man* (New York: McGraw-Hill, 1964), 22, 29.

11. Infrastructure has a long and complex history, one explained in the introduction to this volume. For my purposes here, I consider infrastructure in terms of "pervasive enabling resources in network form," drawn from Geoffrey C. Bowker et al., "Toward Information Infrastructure Studies: Ways of Knowing in a Networked Environment," in *International Handbook of Internet Research*, ed. Jeremy Hunsinger (Philadelphia: Springer, 2010), 98.

12. I am not suggesting that verisimilitude and compression are the only governing metaphors to get us through the story, or that compression should simply replace verisimilitude as a single master narrative. There are certainly others. But there is no one single narrative, and if we don't want to keep repeating the same old stories, we need to learn to tell other ones.

13. Jonathan Sterne, *mp3: The Meaning of a Format* (Durham, N.C.: Duke University Press, 2012), 4–5.

14. The dream that technological media will somehow reproduce the fullness of face-to-face communication has been roundly criticized from poststructuralist and

pragmatist positions over the past fifty years. See Jacques Derrida, *Of Grammatology*, trans. Gayatri Chakravorty Spivak (Baltimore, Md,: Johns Hopkins University Press, 1976); Friedrich Kittler, *Gramophone-Film-Typewriter*, trans. Geoffrey Winthrop-Young and Michael Wutz (Stanford, Calif.: Stanford University Press, 1999); Briankle Chang, *Deconstructing Communication: Representation, Subject, and Economies of Exchange* (Minneapolis: University of Minnesota Press, 1996); John Durham Peters, *Speaking into the Air: A History of the Idea of Communication* (Chicago: University of Chicago Press, 1999). Symbolic interactionists have also questioned the premise that interpersonal communication has the fullness that media theorists sometimes attribute to it. Those who have thought most deeply about interpersonal communication have found it to be fraught with power differences, gaps, and anxieties. Those who have thought most deeply about the senses have found them to filter and transform the outside world in making it available to the perceiving subject. See the following books by Erving Goffman: *The Presentation of Self in Everyday Life* (Garden City, N.Y.: Doubleday, 1959); *Stigma: Notes on the Management of Spoiled Identity* (Englewood Cliffs, N.J.: Prentice-Hall, 1963); *Relations in Public* (New York: Harper and Row, 1971). See also Leslie A. Baxter and Barbara M. Montgomery, *Relating: Dialogues and Dialectics* (New York: Guilford, 1996).

15. Gilbert Simondon, "Technical Mentality," *Parrhesia* 7 (2009): 23.

16. Gilbert Simondon, *On the Mode of Existence of Technical Objects*, trans. Ninian Mellamphy (London, Canada: University of Western Ontario, 1980), 62, emphasis added.

17. Ibid., 47. As Brian Massumi explains, "It has to do with the potential for the oil in the turbine and the water around it to each play multiple roles. The water brings energy to the turbine, but it can also carry heat away from it. The oil carries the heat of the generator to the housing where it can be dissipated by the water, but it also insulates and lubricates the generator, and thanks to the pressure differential between it and the water, prevents infiltration." Brian Massumi et al., "Technical Mentality Revisited: Brian Massumi on Gilbert Simondon," *Parrhesia* 7 (2009): 39.

18. *Oxford English Dictionary*, s.v., "compression"; John R. Taylor, "Limiting Amplifiers," *Communications* (December 1937): 7.

19. Ibid., 10.

20. "You have this relatively new choice: either to discipline the people or to substitute for the unreliable humans a delegated nonhuman character whose only function is to open and close the door." Bruno Latour, "Mixing Humans and Nonhumans Together: The Sociology of a Door-Closer," *Social Problems* 35, no. 1 (1988): 301.

21. Taylor, "Limiting Amplifiers," 10.

22. Ibid.; Karin Bijsterveld, *Mechanical Sound: Technology, Culture, and Public Problems of Noise in the Twentieth Century* (Cambridge, Mass.: MIT Press, 2008), 87; Albin J. Zak, *The Poetics of Rock: Cutting Tracks, Making Records* (Berkeley: University of California Press, 2001); Louise Meintjes, *Sound of Africa! Making Music Zulu in a South African Studio* (Durham, N.C.: Duke University Press, 2003); Jonathan Sterne, "Enemy Voice," *Social Text* 25, no. 3 (2008).

23. Gilles Deleuze, *Foucault*, trans. Seán Hand (Minneapolis: University of Minnesota Press, 1988), 23–44.

24. As the editors to this volume point out, infrastructure is usually hidden from end users. Like other media forms, it works to make itself imperceptible.

25. Simondon, "Technical Mentality," 19.

26. Shannon, C. E. "A Mathematical Theory of Communication," *Bell System Technical Journal* 27, no. 3 (1948): 381.

27. Ibid., 398–99.

28. Ibid.

29. Harold Innis, *The Bias of Communication* (Toronto: University of Toronto Press, 1991), 115–16.

30. Lazlo Solymar, *Getting the Message: A History of Communications* (Oxford: Oxford University Press, 1999), 3–19; Russell W. Burns, *Communications: An International History of the Formative Years* (Herts, UK: Institution of Electrical Engineers, 2004), 5–20.

31. Burns, *Communications*, 47.

32. Ibid., 42.

33. Solymar, *Getting the Message*, 28.

34. See, for example, Daniel Czitrom, *Media and the American Mind: From Morse to McLuhan* (Chapel Hill: University of North Carolina Press, 1982); James Carey, *Communication as Culture* (Boston: Unwin Hyman, 1988); Menachem Blondheim, *News over the Wires: The Telegraph and the Flow of Public Information in America, 1844–1897* (Cambridge, Mass.: Harvard University Press, 1994); Richard John, *Network Nation: Inventing American Telecommunications* (Cambridge, Mass.: Harvard University Press, 2010).

35. Charles (as C.M.) Morrison, "An Expeditious Method of Conveying Intelligence," *Scots Magazine* 15 (1753).

36. Schilling's experimental telegraph (1820–1830) used five discs that each had two positions, allowing it to represent all letters of the alphabet through a binary code. Burns, *Communications*, 68.

37. Ibid., 76, 83–84.

38. Jonathan Sterne, *The Audible Past: Cultural Origins of Sound Reproduction* (Durham, N.C.: Duke University Press, 2003), 147–51.

39. N. Katherine Hayles, *How We Think: Digital Media and the Contemporary Technogenesis* (Chicago: University of Chicago Press, 2012), 124.

40. Ibid., 131–32.

41. Ibid., 151.

42. Ibid, 216.

43. Hayles is drawing on philosopher of mind Andy Clark with the coinage "continuous reciprocal causality," though she also refers to Simondon on several occasions. Ibid., 86. See also Andy Clark, "Time and Mind," *Journal of Philosophy* 40, no. 7 (1998).

44. Richard Ohmann, *Selling Culture: Magazines, Markets and Class at the Turn of the Century* (New York: Verso, 1996), 21, 234–39, quote at 34. See also Ulrich Keller,

"Photojournalism around 1900: The Instiutionalization of a Mass Medium," in *Shadow and Substance: Essays on the History of Photography in Honor of Heinz K. Henisch*, ed. Kathleen Collins (Bloomfield Hills, Mich.: Amorphous, 1990); Kevin G. Barnhurst and John C. Nerone, *The Form of News: A History* (New York: Guilford, 2001), 19.

45. Mara Mills, "The Dead Room: Deafness and Communication Engineering," PhD diss., Harvard University, 2008, 83–153. "Deafening: Noise and the Engineering of Communication in the Telephone System," *Grey Room* 43 (2011); Sterne, *MP3*, 19, 32–60.

46. Jonathan Sterne and Dylan Mulvin, "The Low Acuity for Blue: Perceptual Technics and American Color Television," *Journal of Visual Culture* 13, no. 2 (August 2014): 188–98; Dylan Mulvin and Jonathan Sterne, "Scenes from an Imaginary Country: Test Images and the American Color Television Standard," *Television and New Media* (forthcoming).

47. Raymond Williams, *Culture and Society, 1780–1950* (New York: Columbia University Press, 1958); *The Long Revolution* (London: Chatto and Windus, 1961).

48. John Mowitt, "The Sound of Music in the Era of Its Electronic Reproducibility," in *Music and Society: The Politics of Composition, Performance and Reception*, ed. Richard Leppert and Susan McClary (New York: Cambridge University Press, 1987); Peter Krapp, *Noise Channels* (Minneapolis: University of Minnesota Press, 2011); Michel Serres, *The Parasite*, trans. Schehr Lawrence R (Baltimore, Md.: Johns Hopkins University Press, 1982); Greg Hainge, *Noise Matters: Toward an Ontology of Noise* (New York: Bloomsbury Academic, 2013); David Harvey, *The Condition of Postmodernity: An Enquiry into the Origins of Cultural Change* (Cambridge: Blackwell, 1989).

CHAPTER 2

Fixed Flow

Undersea Cables as Media Infrastructure

NICOLE STAROSIELSKI

With each wave of technological development, the media landscape appears less wired. Mobile phones, tablets, and laptops enable users to access content in an array of environments, appearing to connect only intermittently to an electric or communications grid. Wireless devices are used to control media technologies at a distance, whether radios, television screens, or video games. Contact with digital systems today is marked by what Adrian Mackenzie describes as "wirelessness," an experience of being entangled with wireless technologies and services at many different places and times, along with the indistinct sensations of interference and weak connection they generate.[1] This experience even extends to the fringes of our global networks, where some newly connected locations have "leap-frogged" traditional land-line telecommunications infrastructure. If there is a spatiality to this imagination, it is one that positions signal traffic as moving up into the atmosphere and across the airwaves to visible cell towers, antennas, and satellite dishes. Cloud computing—the name given to distribution systems that store content out on the network rather than on a personal computer—draws from this aerial imagination to depict everything from storage programs to music delivery platforms as hovering above the fixed realities of the material world.

This proliferation of wireless media technologies is grounded by a large mass of cable systems. Buried under soil and pavement, snaking along the bottom of the ocean, enclosed in industrial parks and office buildings, and secluded in rural areas, the majority of Internet traffic passes through their circuits. In the United States, every time users search for information using Google, post a picture to Instagram, or dial a number on Skype, they activate a part of this subterranean and subaquatic infrastructure. These digital media companies run their platforms from large servers, often kept in remote sites. From users' laptops, the signals might move to a router, out into a local urban network, switching to a different network at an Internet exchange, speeding along a long-haul underground terrestrial backbone, through a cable station on the coast, into an undersea cable (all the while being boosted intermittently by repeaters), through another cable station, to a warehouse-like data center, and back again. Even if users ultimately encounter content wirelessly, it is often only in the last few hops between laptops and routers, or between cell towers and mobile phones that signals are freed from the grid.

Of this wired landscape, undersea cables constitute the section that makes the Internet a global phenomenon. Almost 100 percent of transoceanic Internet traffic is carried via fiber-optic undersea cables and, at times, is transmitted this way when it is moving between locations on the same continent. A number of authors—from science fiction writer Neal Stephenson to technology journalist Andrew Blum—have written about the construction of undersea systems, but the coverage of cable networks has been sporadic, and they remain, for the most part, absent from our everyday technological imagination.[2] Although undersea fiber has connected continents from the late 1980s onward, only in the last decade has it drawn substantial attention from researchers. Geographers, including Barney Warf, Edward Malecki, and Hu Wei, have documented the transition from satellite to fiber-optic dominance, brought about in the 1990s by the higher capacity and cost-effectiveness of cable systems.[3] Policy researchers have written critical reports about the vulnerability of submarine networks to major disasters, especially after cable-disrupting events such as the 2006 Hengchun earthquake.[4] Historians, inspired by the twenty-first-century turn to a technology with nineteenth-century roots, have reflected on the parallels and divergences between successive waves of cable systems.[5] In media and communication studies there has been a turn toward infrastructure; fiber-optic systems, a key component of our information distribution, have emerged as part of that discussion.[6]

Undersea cables, however, remain difficult to connect to the questions that have traditionally animated media studies about the production of sounds and images, the formal characteristics of texts, and the relationship between media content and culture. Cable researchers have documented large-scale transitions within the telecommunications industry, the influence of key institutional actors, and the relationship between fiber-optic networks and global capitalism, but they have seldom connected cable networks to the everyday politics and practices of media distribution—this is to say, undersea cables are rarely investigated as a *media* infrastructure. This might be in part because they are one of the deepest strata of the Internet—laid on the bottom of the seafloor, they appear relatively distant from the user interface and embodied encounters with media content. Signals must traverse several layers of technological systems—down through computer processors, hard drives, and routing protocols—before reaching undersea networks. Beyond this are the numerous systems on which cables themselves depend, from air-conditioning technologies to the labor of maintenance workers. Still further below are fossil fuels that provide energy for cooling and topographic forms that permit the building of transportation networks: nature, as Paul Edwards has observed, is "in some sense the ultimate infrastructure."[7]

What does it mean to describe any of these interconnected layers of digital systems—from routing protocols, to cable networks, to fossil fuels—as a media infrastructure? How can this description be made meaningful to the analysis of such distant circulations of content and made relevant to media and communications research? In this chapter, drawing from work by Susan Leigh Star and Karen Ruhleder, I offer a framework to conceptualize the relationships between cascading layers of interlinked technological, social, and environmental systems and the distribution of signals that they ultimately support. Star and Ruhleder observe in their seminal essay, "Steps Toward an Ecology of Infrastructure," that the concept of infrastructure is fundamentally both relational and contextual. They argue that a technical system becomes an infrastructure only through its use: what exists for one person as a media infrastructure might be for others an impediment to media circulation. They suggest that we ask "*when*—not what—is an infrastructure."[8] To make a parallel claim in a media-studies context: to consider technical, social, and natural systems as media infrastructures entails understanding *when* particular systems are infrastructural for mediation and how these systems differentially shape the dissemination of media culture. Analyzing cables as media infrastructures involves articulating how they invisibly contort the conditions of possibility, geographic dispersion, and cultural perception of media signals. This approach blurs any preexisting distinction between media infrastructures and other kinds of infrastructures such as shipping lines, roads, and rivers, and instead considers how flows of audiovisual content and technical, social, and natural systems are always constituted in relation to each other. At the same time, it is important to recognize limits to such infrastructural connections: though undersea cables support the circulation of media, they are not infrastructures for all kinds of media at all times, and this is especially true when we examine their diverse global contexts.

To elaborate this approach, in this chapter I chart four ways that cables emerge as media infrastructures, influential for and perceptible in audiovisual circulations. First, cables function as a resource, both real and imagined, for mediation. Fiber-optic networks are often intertwined with speculations about and economic investments in media projects and systems. Their presence serves as a rationale for the development of new media industries and practices, and, in turn, these industries and practices are called on as imagined markets in proposals for new cable projects. It is this insular feedback loop—wherein cables are seen as resources for audiovisual media and in turn media are projected as resources for cables—that enables the speculative development of large-scale infrastructural projects in the absence of any actual circulation. The second way that cables inflect media distribution is in their alteration of the temporality of information exchange, not only for the media industries, but also for individual users. As they contort the time it takes for signals to transit between locations, cables have a tangible effect on media practices such as online gaming that depend on high-speed and high-bandwidth transmission, producing what I call an "aesthetics of lag" when content is not efficiently transmitted. Third, cables implicate users within new and unseen structures of power. Depending on their geography, cables might increase the susceptibility of media to censorship or surveillance. Cable routes are places where media systems can be disrupted, where infrastructures can become entangled in local politics, and where concerns about privacy play out. Rather than extending uniformly across space, cables have often remained embedded in existing geographies, and their effects on media industries, user experiences, and the politics of circulation occur unevenly around the world. This observation brings us to the chapter's final point: cables can perpetuate imbalances in media production and consumption, an inequality that becomes most apparent in the differing cost of media access.

The examples I describe below reveal that, despite their apparent distance from everyday media practices, cables are thoroughly intertwined with media production and consumption. Yet cable systems do not simply determine the movements of media but, rather, are situated as part of a feedback loop. Network infrastructure emerges through users' everyday practices. Depending on where they are in the world and the platform they are operating from, users activate and inhabit different slices of this wired infrastructure. In some locations the content they are seeking might be stored locally, and data has to travel only a short way between its origin and destination. Other content might have to circumnavigate the globe. In their engagements with certain forms of media, and in their differential activation of infrastructure, users are unknowingly entangled in specific kinds of infrastructural development. Through this process, changes in media practices aggregate to alter the economics, practices, temporality, and geography of undersea cable systems.

Although this chapter temporarily extricates cables from the broader networks of telecommunications and computation, it offers a framework for the consideration of all communications technologies, and even human, non-human, and natural environments more broadly, as infrastructures for audiovisual media. Technologies and environments might be imagined as resources for the generation and transmission of sounds and images, even when their capacity is not utilized. They alter the speed of distribution and, as a result, users' understanding of media temporality. They can be potential sites for media disruption, censorship, and political intervention. Though many of these technologies and environments are hidden from view, they formatively structure and imperceptibly affect our experiences of media. In turn, users activate networks in partial and unpredictable ways but at the same time can engage in a politics of infrastructure through their everyday media practices. As the following examples reveal, conceptualizing systems such as undersea cables as *media infrastructures* can help scholars to better account for investments, aesthetics, and inequalities in media, and to understand the interfaces between our experience of digital content and the technologies that make its circulation possible.

Infrastructure as Resource

Like radio towers, television channels, and postal trucks, undersea cables are a technological system that affords capacity for the distribution of words, images, sounds, and other audiovisual material. Historically these technologies have altered and expanded the possible ways that media can be transmitted. For example, in the nineteenth-century telegraph network, undersea links were a key transmission device for global news agencies such as Reuters and the Associated Press. These agencies' use of cables drastically changed the experience of readers who had previously waited months for news from abroad. The economics of the new system, however, meant that sending telegrams remained too expensive for most everyday communication. As Dwayne Winseck and Robert Pike have argued, the development of the cable network helped give rise to the first global media system, one characterized by tight connections between the companies that controlled network infrastructure and the companies that circulated content using cables.[9] In their first stage of infrastructural development, undersea systems were configured as a resource facilitating the production and consumption of news media.

Today's fiber-optic cables still play an important role in transmitting news. They enable the cheap bandwidth that makes practical the worldwide dissemination of Twitter updates and blog posts detailing global events in real time, alongside the online content of traditional news outlets. They also significantly shape the possibilities for bandwidth-intensive networked audiovisual media industries, including film and television. Immediate and reliable Internet access is critical, for example, for on-location film productions that require real-time communication. Cities that aspire to support intensive international media collaborations likewise need fiber-optic infrastructure. Cables are especially important for work that depends on digitization, including digital animation and special effects, since these high-speed links render the frequent and instantaneous sharing of high-resolution video effective and economical. In her study of *The Lord of the Rings* trilogy, Kristen Thompson discusses the importance of the "Fatpipe," a dedicated network that connected Wellington to London via both undersea cables and terrestrial lines. While the film production's domestic communications were transmitted via satellite, cables were needed to send large amounts of footage to London. The crew on both ends also used the system for videoconferencing: cables made it possible for them to simultaneously view and comment on footage.[10] Distribution resources, as Nadia Bozak argues, critically inflect the production of the image, constituting its underlying "physical and biophysical makeup"[11] and generating specific kinds of production potentials.

Once established, this physical makeup can become a catalyst for other kinds of media flows. Cables are a "precious manmade resource," according to those in the cable industry, and enhance capacities for transnational media collaboration.[12] The Fatpipe became a model and resource that was mobilized for subsequent productions, including the effects on *I, Robot* (Proyas, 2004) and *Avatar* (Cameron, 2009). In some places, nations view fiber networks as a resource to be leveraged toward the creation of new media industries and practices. After investing in a cable system (the same one that carried *The Lord of the Rings*), the government of Fiji looked for a way to generate traffic and pay off their investment. They developed information and communication technology parks, call centers, and a new initiative, "Bulawood: The Hollywood of the South Pacific," all of which depended on the cable for transnational exchange and featured its capacity in their marketing.[13]

The relationship between the establishment of a cable system and the growth of the media industries is not a simple circuit. Shifts in the cable industry, especially economic shifts, affect the way that these systems can be used as media infrastructures. The emergence of the Fatpipe itself reflected recent changes in the undersea cable world. As Thompson observes, at the time of *The Lord of the Rings* production the "Internet bubble had recently burst, and much of the broadband capacity on the Internet was suddenly lying idle."[14] As a result, the filmmakers were able to establish the network at relatively low cost. In addition, the same cables used by media industries can undermine their efforts, since they often make it easier to circulate pirated material. The relative accessibility of the Internet via the Southern Cross cable, for example, invigorated Fiji's illegal DVD distribution networks, which in turn compromised efforts to expand commercially profitable domestic film distribution.[15] While their effects on the mediasphere are quite varied—in some places facilitating large-scale digital media development, and in other cases compromising it—cables serve as an imagined resource, capacity "lying idle," which many believe must be channeled to support media industries.

Within the undersea cable industry, media consumption and production has in turn been imagined as a resource: it is seen as a key driver of traffic, a source of flow that will generate a sizeable stream of income. In the late 1990s sales pitches of the cable companies, media was touted as the reason to invest in an undersea network. A team from Alcatel Submarine Networks argued that "even if the growth in computers in the workplace eventually starts to level off, the increasingly sophisticated software—including all flavours of real-time collaborative work and multimedia applications—run on these computers will continue to generate expanding bandwidth requirements."[16] This figuring of media as both stimulus and resource for network expansion continues to permeate the industry. William C. Marra, the CEO of a company planning to lay a new transatlantic network, reported that they anticipated "explosive growth" in traffic due to the "continued market expansion of media."[17] Elsewhere, video has been described as "fueling" the demand for undersea systems—a parallel that links media to the imagined resource economies of oil, gas, and timber.[18] High-definition television is seen as an important development in particular because of the 40 percent to 70 percent more bandwidth it requires.[19] The move to "media-rich content," including streaming video, online gaming, and other forms of cloud computing, rather than voice traffic remains a rationale for the construction of new networks.[20] Cable systems are pitched to investors and often funded on the basis of such speculations about a future media environment.

Anticipating such media developments, some cable companies design specialized services to appeal to the media industries and transnational collaborations. To send content around the world for *The Lord of the Rings* productions,

Telecom New Zealand developed a specialized Film Net service.[21] Hibernia Networks, which operates a transatlantic cable system, has a media division that facilitates live video transmissions for large media organizations. They have even drawn staff from the television industry to operate their Television Operating Center, built to manage the distribution of broadcasts around the world and to ensure video would be transmitted with low latency (in other words, no delay), low "jitter," and high-resolution.[22] Cables are sometimes laid with particular media events in mind: for example, the Trans-Pacific Express cable, extending between the United States and northern China, was timed to coincide with the Beijing Olympics. Companies that have historically focused on the production, categorization, and distribution of content have even begun to invest directly in cable infrastructure: Google recently laid a new transpacific cable, Unity, between Los Angeles and Chikura, Japan.

What results is a loop: media industries, governments, and other organizations see undersea cable networks as a resource to be leveraged, an open channel with unused capacity much like a highway or a set of train tracks. The presence of fiber-optic cables forms a rationale for media development: because there is potential infrastructural connectivity, expanding a media sector is a way to capitalize on existing resources. On the other hand, cable companies see the media circulations of users, of cloud computing companies, and of various other industries as a resource, a set of unruly flows that can be channeled and made profitable, much like a river or an oil reserve. Harnessing these flows will in turn, they believe, generate demand for additional networks. While cable industry rhetoric evokes the metaphors of automobility—or containerized movement—it also conjures up a scenario in which cars stand idle with no roads available, an imagination that helps to justify the building of more cables. This combination, in which cables are seen as resources for media and media are seen as a resource for cables, is key to the actual funding mechanism for today's network infrastructure.

Infrastructure and Media Temporality

While fiber-optic cables shape large-scale conditions of possibility for the institutional and industrial production of media, they also influence everyday digital media experiences. Cables and network infrastructure broadly affect media circulations by regulating the speed at which media is transmitted between locations. Fiber-optic cables enable faster signal exchange than wireless modes of communication: it takes about one-eighth the time for a signal to travel by cable between New York and London as it does by satellite. This difference matters for some communications practices more than others. A delay has less effect on most users' experience of sending and receiving email than it does on video transmission, which might be distorted by a lag in the image or audio. As a result, digital media access can be quicker and easier for users in areas equipped with fiber-optic cables and which have efficient network routing.

For applications unaffected by time delay, such as email, the geographic organization of networks does not appear to make a difference, nor is it perceptible in users' media exchanges. For time-sensitive applications and practices, however, the geography and organization of infrastructure can critically affect one's media experiences. This has been well documented by the players of massive multiplayer online games. In his study of *Counter-Strike*, a networked first-person shooter game, anthropologist Graham Candy observes the critical and perceptible role material infrastructures such as undersea cables and network servers have in shaping gameplay. For *Counter-Strike*'s players, finding a high-quality and proximate gaming server decreases any delay in their navigation of the game, allows them to play in near-real time, and gives them a tangible advantage over other teams.[23] As a result, despite their ability to join forces with other players around the world, many instead choose to play on servers closer to home. Using a distant server, being forced across excess cable links and network nodes, in contrast, gives the sensation of "lag," what Candy describes as a "visceral, emotional and physical reaction" of being slowed down, and users who "are literally feeling the bricolage of infrastructures" become angry that they cannot experience the game as intended.[24]

Lag is not only experienced by gamers but appears across media platforms, especially on those hosting digital video. Frozen images, scrambled representations, out-of-sync sound, and the seemingly endless buffering of streaming television, alongside low-resolution, compressed versions are familiar parts of the process of watching video online. Paul Benzon observes that such failures "might characterize the experience of digital video consumption as much as its promised purchase upon the cinematic."[25] These failures are often due to the efforts of users to access *too much* content, an attempt to exceed the capacity of the system. Such aesthetics of distribution are not limited to the transmission of media via cables. As Lucas Hilderbrand argues, the copying and recopying of bootleg VHS tapes, which led to their deterioration over time, marked them with a distortion and fuzziness that he terms the "aesthetics of access."[26] While the bootleg aesthetics of videotape are produced by the materiality of storage technology and playback devices, in the case of online video what I call the "aesthetics of lag" is inscribed on the image by the materiality of network infrastructure. These images register the inability of networks (and the companies that run them) to appropriately manage the distribution of content across transmissions lines, servers, and other infrastructural components. In these moments—of scrambled calls and distorted images—we can perceive the inability of signals to properly transit media infrastructure. In contrast, the low-fi aesthetics of compression are marked by the *proper* management of content for the transmissions lines, the decreasing of signal size in order to fit media infrastructure. As was true for videotape's aesthetics of access, Jonathan Sterne observes that the aesthetics of compression have even become pleasurable for some audiences.[27]

The aesthetics of lag represent one way in which media consumers come into contact with infrastructure, though this encounter tells them little about what infrastructures they are traversing or where their signals extend. On one hand, lag and distortion are not merely products of distance. One does not simply experience lag when accessing content that is far away. The long haul between continents is a speedy trip relative to the time it can take to move through an Internet exchange or a local network. In describing the microseconds it takes a signal to transit a router, Andrew Blum writes:

> compared with the amount of time it takes a bit to cross the continental United States . . . that time spent crossing the router was an eternity. It was like walking ten minutes to the post office only to wait in line for seven days, around the clock . . . powerful though they may be, [routers] were the traffic-clogged cities on a journey across the open net.[28]

Content often experiences relatively more "traffic" at off ramps and interchange points. In some places, speed is as much of an economic problem as a technological one, since companies that manage signal exchange have a financial incentive to squeeze the most traffic through the fewest possible circuits. Internet service providers and state authorities might even engage in "bandwidth throttling," the intentional slowing down of signal exchange in order to manage high-bandwidth activities on the network.[29] Due to the variability of these infrastructural geographies, it might take longer to transit some cities than to cross an entire ocean. Therefore, even though cables might visibly shape the temporality of media content, the traces that they leave rarely give users any clues about the composition of media infrastructure.

Infrastructure and Media Disruption

The material geographies of cable infrastructures affect the trajectories of media content not only in determining when it slows down or speeds up but by establishing a matrix of locales in which circulation can be interrupted. Such was the case when, after several cable-cuts off the coast of Egypt, Internet connectivity was substantially reduced in the Middle East. Network disruptions are not an infrequent occurrence: undersea cables break every three days.[30] Many of these are due to human interference, ranging from trawlers that drag nets over the ocean floor to local infrastructural projects that dig up cables on land. Cable theft is also a problem, far more so for terrestrial than subsea links, which are largely protected by the ocean above them.[31] *Capacity Magazine* reported there were approximately ten cable thefts a day across different industries in Germany, ultimately costing the companies who owned them millions of dollars.[32] Most of these breakages cause no perceptible difference in our access to media. Nonetheless, the necessity of repair and maintenance is a cost built into the expense of operating cable systems, and across the network the threat of potential disconnection remains. Media companies who rely on cable infrastructure have a major stake in its smooth operation, and their own reliability is jeopardized if cables do not remain secure.

Cable routes are locations where media and communications are not only materially disrupted but can be actively surveilled or censored. British control of telegraph networks in the late nineteenth century gave the country the ability to intercept and censor messages, an activity that manifested especially during wartime.[33] Similarly, Alfred W. McCoy has documented how U.S. control over telegraph networks in the Philippines helped the United States to develop as a surveillance state.[34] Cable surveillance capacities are still being leveraged today. After the Edward Snowden leak of National Security documents, *The Guardian* published an article about the British Government Communications Headquarters' infrastructural monitoring and drew public attention to the surveillance of data collection at cable stations and other media infrastructure sites.[35] The National Security Agency's Upstream program also collects information as it passes through fiber-optic links.[36] Because cables extend through national territories, the media that transit them are susceptible to the monitoring capabilities and infrastructural power of these nations—even if content is not sent or received from there.

This is complicated by the fact that media traffic does not always move in a direct geographic route between locations and instead assumes a twisted economic geography. Signal traffic often follows the least expensive rather than the quickest route. In some locations it is less expensive to buy a direct circuit to somewhere with "cheap" Internet instead of buying Internet access out of one's own country, a scenario called "pipe and port" in the industry.[37] A company in Hong Kong might choose to pay $70,000 per month locally to access the Internet, or they might choose to purchase a dedicated circuit to Los Angeles ($43,800 per month) and then access the Internet there ($19,300 per month). This would make it slightly less expensive for Hong Kong companies to route all Internet traffic across the ocean to Los Angeles before going anywhere else in the world. In many places, especially in South America, the Middle East, and Africa, it is much cheaper to route Internet traffic through another country; therefore, when media is transmitted between two locations in a single country, it may cross into foreign territory. The research firm Telegeography recently reported that in São Paulo, Brazil, between 2009 and 2012, the local cost to access the Internet was 150 percent that of running traffic through Miami.[38] Essentially, this means that media exchange routed via the Internet in São Paulo, even when directed to other proximate locations, might be moving along undersea cables to Miami and back again. If the NSA is monitoring traffic in Miami, they may very well intercept email sent between two locations in São Paulo. Since more than 80 percent of the Middle East and Africa's traffic is exchanged in Europe, these transmissions also remain susceptible to remote foreign monitoring or intervention.[39]

Given the lack of information circulated about monitoring practices, cables are sites where concerns about information and media censorship emerge. For example, in his pitch for a trans-Arctic cable, CEO Douglas Cunningham noted that the Arctic Fibre system would have circuits directly between London to Japan but would not land on U.S. territory. This option might be of special interest, he suggested, for Asian or Middle Eastern telecommunications carriers that did not want their content subject to U.S. laws. Anxieties about interception surface not only around the routes and geographies of transmission but also in the public discussion of network materials and supplies. The construction of transoceanic systems is currently dominated by two companies—TE Subcom, an American company, and Alcatel-Lucent, a French company. For major intercontinental systems, any cable project will likely contract with one of the two. In 2008, the Chinese company Huawei Marine was launched and has since provided undersea cable for small-scale networks in the Mediterranean, off the coast of South America, and between Indonesian islands. After Huawei Marine signed a contract to build a prominent connection between New York and London, one of the most heavily trafficked routes in the world, the United States House Intelligence Committee released a report warning of the risks in using a Chinese supplier, suggesting that Chinese-made equipment could be used to tap content.[40] Hibernia Networks subsequently halted their work on the cable system and eventually moved to the American vendor.[41] As described in the first section, even the imagined geographies of cable infrastructure can affect the circulation of media content.

Cable Geographies and the Construction of Inequality

Every version of cable technology differentially affects the mediascape, privileging certain forms of content, access, and geographic dispersion over others. While early telegraph cables had a revolutionary influence on the distribution of news media, especially in the West, these networks did not support the dissemination of photography. While voices could be reduced to scripted dialogue and converted into Morse code, the acoustic voice could not be transmitted via cable, nor could music, audio recordings, or moving images once they emerged at the turn of the century. Even though phototelegraphy and undersea telephone cables were later introduced, the initial telegraph networks had the most widespread effect on practices that could be encoded into a text of dots and dashes and therefore were used disproportionately by the news industries that could afford them. In the process, undersea telegraph cables sped up the movement of news but simultaneously transformed the perception of what "fast" was, reducing the relative speed of visual culture and noncabled media. Media infrastructures accelerate the transmissions of certain actors and industries but always do so at the cost of reducing the relative speed of others.

Cables' transformation of media's temporality also occurs unequally across the network. In many places, those who can afford to—whether filmmakers, gamers, or bankers—can pay for a higher-quality network experience and "low-latency" routes. Peter Jackson's signals routed through the Fatpipe were not delayed as they crossed the Pacific, but other local users received sluggish Internet speeds (in many cases due to crowding on domestic systems rather than the distance of the transoceanic links). High-frequency traders on global stock markets use computer algorithms to take advantage of the slight price changes in different locations, secure trades at slightly quicker rates, and exploit short cable paths for profit. Any given company's, nation's, or individual's ability to mobilize cables as a resource, avoid lag and compression, bypass surveillance, and pay a discounted price for access depends in part on their geographic and socioeconomic position. U.S. consumers worry less about such inequalities, since most of the content they wish to use is relatively accessible, stored domestically, and linked to the user via diverse cable routes. In many places, media content speeds across the ocean only to slow or stop at the shore, making access difficult and expensive for landlocked locales. As a result, communities on the periphery of current networks face a disadvantage in a cabled era and remain more vulnerable to disconnection or monitoring.

These inequalities are not likely to be overcome in the near future. Given the scarcity and expense of undersea cables, not every location will receive one, and many will be fortunate to set up two. Many networks cost hundreds of millions of dollars; some cost over a billion. Moreover, every signal sent costs money and consumes energy. It is expensive to maintain the computers that route the signals; upgrades need to be conducted and operations staff must be hired. It costs money to power these systems, to push signals under the bottom of the ocean, and to cool off the cable stations with air conditioning. Funding is required to pay cable ships and crews to stand by in case a cable is broken and needs to be fixed. The expense of transiting such systems, given the unequal geographies they extend through, varies widely. For example, Malecki and Wei report that in 2005, the median revenue of a circuit from the United States to Nigeria and Vietnam was $44,000, compared with $21 from the United States and the United Kingdom. More recently, in 2012 it cost $19,300 per month to access 10GB of Internet content per second in Los Angeles, and more than three times as much in Hong Kong. Media simply costs more to send to and through different parts of the world. As they seek to circulate content across these differential topographies, some media producers and consumers thus occupy positions of privilege while others face disproportionate challenges.

Activating the Network

When circulating media through networked systems, users occupy a particular slice of infrastructure. Their decisions about how and what to produce, consume, and distribute implicitly support different modes of infrastructure development. To communicate with each other, users in countries with advanced digital infrastructure have an array of options: SMS, email, telephony (mobile or land line), and video. Each of these takes up a certain amount of space on undersea networks and reflects particular modes of technical and industrial organization, and this can make a significant difference in cases where bandwidth is limited or expensive. To choose to send a short text message is to be frugal with bandwidth, whereas to transmit video is to be excessive. While the economic and technological relationships between media practices and the development of actual cable networks is complex, if enough users send text messages instead of supporting high-bandwidth media, this would alter the economic model of the cable industry, which is built on projecting media as an ever-expanding resource to be capitalized on. This could in turn make it more difficult for these companies to secure loans to build infrastructure. On the other hand, if users included high-definition video in all communications, this would instead precipitate the speculative futures that private cable companies profit from. As users determine how to consume media and what they are willing to pay, they support and inhibit particular cycles of infrastructural development.

In their movements through content and platforms across different parts of the network, users also activate different media geographies, participating in the unequal distribution of signal traffic. This is perhaps most perceptible in the shift to cloud computing—the move toward a reliance on media stored elsewhere. Keeping one's data in the cloud entails an increasing reliance on undersea cables (and other international communications infrastructure) to connect to content that may have previously been locally available. For example, by using Dropbox to store files instead of keeping them on a local hard drive, one might very well become susceptible to the surveillance of foreign countries, their economic decisions, and policies about privacy (or lack thereof). There are very few ways to be able to determine one's actual entanglements with such systems: there is no information given about the specific data centers one's media is held in or the cables it transits, never mind the ways in which it might be surveilled. The user is not a rational agent who can locate herself in relation to such infrastructures; rather, she is a posthuman subject that extends across the network in multiple, unpredictable ways, intertwined with developments that are beyond any individual's knowledge or control.

Although this chapter has focused on undersea cables, the relationships described here hold true for a range of digital media infrastructures. Internet exchanges, the place where media signals are transferred between networks, have become rationales for media development. The massive expansion of data centers has been shaped by changes in user practices. Terrestrial links, which include both national long-haul networks as well as local city networks, are sites where cables can be disrupted. And it is the server, rather than the cable system, that gamers want to be closest to in order to decrease lag. Cable systems are partial infrastructures, and they exist in an ecology that is social and technical, human and nonhuman. To grasp how digital media circulations are shaped by and inflect Internet infrastructure, undersea systems must be considered in relation to these other network components as well as the protocols that facilitate movement across them.

Analyzing undersea cables as media infrastructure draws our attention to the ways that seemingly nebulous digital circulations are anchored in material coordinates. By following the routes of our transmissions, we can understand how cables are viewed as and transformed into resources for media industries (or conversely, do not become resources): cable infrastructure both reflects investments in particular sites and increases these locations' capacity for flow. We might also better conceptualize the experience of temporality on networks in different geographic locales alongside their variable susceptibility to disruption, censorship, and interference by different forces and actors, whether the laws of countries or the technological protocols of corporations. Finally, we might better gauge how individual and collective media use paves and occupies particular pathways for distribution, generates economic circulations for some companies over others, and conditions the access of all users. An analysis of cables as media infrastructure ultimately connects the physical dimensions of these network technologies to the broader dissemination of media cultures.

Notes

1. Adrian Mackenzie, *Wirelessness: Radical Empiricism in Network Cultures* (Cambridge, Mass.: MIT Press, 2010), 6.

2. Jeff Hecht, *City of Light: The Story of Fiber Optics* (New York: Oxford University Press, 1999); David C. Chaffee, *Building the Global Fiber-Optics Superhighway* (New York: Kluwer Academic, 2001); Andrew Blum, *Tubes: A Journey to the Center of the Internet* (New York: HarperCollins, 2012).

3. Barney Warf, "International Competition between Satellite and Fiber Optic Carriers: A Geographic Perspective," *Professional Geographer* 58, no. 1 (2006): 1–11; Edward J. Malecki and Hu Wei, "A Wired World: The Evolving Geography of Submarine Cables and the Shift to Asia," *Annals of the Association of American Geographers* 99 (2009): 360–82.

4. Michael Sechrist, "Cyberspace in Deep Water: Protecting the Arteries of the Internet," *Harvard Kennedy School Review* 10 (2009–10): 41.

5. Bernard Finn and Daqing Yang, eds., *Communications under the Seas: The Evolving Cable Network and its Implications* (Cambridge, Mass.: MIT Press, 2009); Dwayne R. Winseck and Robert M. Pike, *Communication and Empire: Media, Markets, and Globalization, 1860–1930* (Durham, N.C.: Duke University Press, 2007).

6. Wendy Chun, *Control and Freedom: Power and Paranoia in the Age of Fiber Optics* (Cambridge, Mass.: MIT Press, 2006); Jennifer Holt and Kevin Sanson, eds., *Connected Viewing: Selling, Streaming, and Sharing Media in the Digital Era* (New York: Routledge, 2014); Tung-Hui Hu, *Cloud: A Pre-History* (Cambridge, Mass.: MIT Press, forthcoming [2015]).

7. Paul N. Edwards, "Infrastructure and Modernity: Force, Time, and Social Organization in the History of Sociotechnical Systems," in *Modernity and Technology*, ed. Thomas J. Misa, Philip Brey, and Andrew Feenberg (Cambridge, Mass.: MIT Press, 2003), 196.

8. Susan Leigh Star and Karen Ruhleder, "Steps toward an Ecology of Infrastructure: Design and Access for Large Information Spaces," *Information Systems Research* 7, no. 1 (March 1996), 113.

9. Winseck and Pike, *Communication and Empire*.

10. Although Thompson notes that the network was not "absolutely essential," without it the production might have been delayed and much more expensive. Kristin

Thompson, *The Frodo Franchise:* The Lord of the Rings *and Modern Hollywood* (Berkeley: University of California Press, 2007), 306.

11. Nadia Bozak, *The Cinematic Footprint* (New Brunswick, N.J.: Rutgers University Press, 2012), 8.

12. Brian Lavallée, "Broadband—From Land to Sea to Economic Prosperity," *Submarine Telecoms Forum* 71 (July 2013): 13.

13. See Nicole Starosielski, "When Fiji Is Not an Island: Converging Histories of the South Pacific's (New) Media Capital?" in *Locating Emerging Media*, ed. Ben Aslinger and Germaine Halegoua (New York: Routledge, 2015).

14. Ibid., 304.

15. Nicole Starosielski "Things and Movies: DVD Store Culture in Fiji," *Media Fields Journal* 1 (2010): 1–10.

16. B. Le Mouel, S. R. Barnes, and P. M. Glbla, "From Point-to-Point Links to Networks—Benefits to Customers and Technical Solutions," *Proceedings of the 1997 SubOptic Conference*, San Francisco, 1997, 304.

17. "Emerald Express Cable to Land in NY via AT&T," *Capacity Magazine*, January 23, 2013.

18. Alan Mauldin, "Fueling Subsea Bandwidth Demand," *Submarine Telecoms Forum* 28 (2006): 28–30.

19. "Preparing for a 3D future," *Capacity Magazine*, February 15, 2011.

20. "Future of the Latin American Telecoms Market," *Capacity Magazine*, January 15, 2011.

21. Peter Griffin, "Telecom Looking to Keep Film Net in the Picture," *New Zealand Herald*, November 25, 2003.

22. See the "Global Media Events" section on Hibernia Atlantic's website: http://www.hibernianetworks.com/solutions.

23. Graham Candy, "In Video Games We Trust: High-Speed Sociality in the 21st Century," *Fast Capitalism* 9, no. 1 (2012).

24. Ibid. See also Hollis Griffin, "Liveness with a Lag: Temporality and Streaming Television," *Antenna*, August 9, 2012.

25. Paul Benzon, "Bootleg Paratextuality and Digital Temporality: Towards an Alternate Present of the DVD," *Narrative* 21, no. 1 (January 2013): 96.

26. Lucas Hilderbrand, *Inherent Vice: Bootleg Histories of Videotape* (Durham, N.C.: Duke University Press, 2009).

27. Jonathan Sterne, *MP3: The Meaning of a Format* (Durham, N.C.: Duke University Press, 2012), 6.

28. Blum, *Tubes*, 161.

29. *"Comcast Corporation vs. Federal Communications Commission and United States of America,"* No. 08-1291, U.S. Court of Appeals (April 6, 2010).

30. Stephen Beckert, quoted in Ryan Singtel, "Cable Cut Fever Grips the Web," *Wired*, February 6, 2008.

31. Tim Phillips, "Fibre-Optic Cable Cuts," *Capacity Magazine*, May 16, 2011.

32. Tim Phillips, "Copper Theives Caught Red Handed," *Capacity Magazine*, July 9, 2013.

33. Daniel R. Headrick, "Radio Versus Cable: International Telecommunications before Satellites," in *Beyond the Ionosphere: Fifty Years of Satellite Communication*, ed. Andrew Butrica (Washington: NASA, 1997), 3–8.

34. Alfred W. McCoy, *Policing America's Empire* (Madison: University of Wisconsin Press, 2009).

35. Ewen MacAskill, Julian Borger, Nick Hopkins, Nick Davies, and James Ball, "GCHQ Taps Fibre-Optic Cables for Secret Access to World's Communications," *Guardian*, June 21, 2013.

36. Craig Timberg, "The NSA Slide You Haven't Seen," *Washington Post*, July 9, 2013.

37. Erik Kreifeldt, "Informing Investment Decisions with Submarine Bandwidth Pricing Analysis," *SubOptic 2013 Conference Proceedings*, April 23, 2013.

38. Ibid.

39. Ibid.

40. Michael S. Schmidt, Keith Bradsher, and Christine Hauser. "U.S. Panel Cites Risks in Chinese Equipment," *New York Times*, October 8, 2012.

41. Laura Hedges, "Hibernia Networks Reveals New Partner for Project Express Cable Deployment," *Capacity Magazine*, May 13, 2013.

CHAPTER 3

"Where the Internet Lives"

Data Centers as Cloud Infrastructure

JENNIFER HOLT AND PATRICK VONDERAU

Emblazoned with the headline "Transparency," Google released dozens of interior and exterior glossy images of their data centers on the company's website in 2012. Inviting the public to "come inside" and "see where the Internet lives," Google proudly announced they would reveal "what we're made of—inside and out" by offering virtual tours through photo galleries of the technology, the people, and the places making up their data centers.[1] Google's tours showed the world a glimpse of these structures with a series of photographs showcasing "the physical Internet," as the site characterized it. The pictures consisted mainly of slick, artful images of buildings, wires, pipes, servers, and dedicated workers who populate the centers.

Apple has also put the infrastructure behind its cloud services on display for the digital audience by featuring a host of infographics, statistics, and polished inside views of the company's "environmentally responsible" data center facilities on its website.[2] Facebook, in turn, features extensive photo and news coverage of its global physical infrastructure on dedicated Facebook pages, while Microsoft presents guided video tours of their server farms for free download on its corporate website.[3] Even smaller data centers like those owned by European Internet service provider Bahnhof AB, located in Sweden, are increasingly on digital exhibit, with their corporate parents offering various images of server racks, cooling and power technology, or even their meeting rooms, all for wide dissemination and republishing.[4] Operating out of a Cold War civil-defense bunker hidden thirty meters under the earth, Bahnhof's "Pionen White Mountains" center (its original wartime codename) offers particularly dramatic sights, complete with German submarine diesel engines for backup, and glowing, windowless rock walls protecting blinking servers stacked underground. Alongside such memorable online representations of server facilities, there is also a recent array of coffee-table books, documentaries, news reports, and other offline forms of photographic evidence that have put data centers on display.[5]

But what are all these images about? What drives this excess of vision that asks us to partake in creating visibility for something that remains essentially invisible? Why do we engage in sharing views of emptied, technified spaces? At first glance, on a surface level, the visible evidence abundantly provided by Google, Apple, or Bahnhof might simply appear as a means of creating a positive public image of the data center business. Given the centralization of data in "the cloud," such pictures persuade users to experience the move of their data to corporate "warehouses" as being safe and secure, by depicting a stable and nonthreatening cloud storage environment.[6] Indeed, the notion of the cloud is a marketing concept that renders the physical, infrastructural realities of remote data storage into a palatable abstraction for those who are using it, consciously or not. In fact, a recent survey of more than one thousand Americans revealed that 95 percent of those who think they are not using the cloud, actually are—whether in the act of shopping, banking, or gaming online, using social networks, streaming media, or storing music/photos/videos online.[7]

However, explaining data-center visibility by pointing to the discourses it shapes, to the metaphorical character of "the cloud," or to the ways the cloud is rendered visible by looking "behind the scenes" of another scale economy can merely be first steps. Looking deeper will lead us to acknowledge that much of what we see in these images is also indicative of the competitive dynamics between Google, Apple, and Facebook. Picturing infrastructure means staking corporate territory, given that this infrastructure as well as the software or services it makes accessible are often proprietary and subject to disputes over interoperability issues.[8]

Following this line of thought, we might still take a further step and start observing the rather intense "technological dramas"[9] playing out in the imagery of digital infrastructure. Google and Bahnhof offer especially pertinent examples of what Langdon Winner called the "politics of artifacts": the way working systems choreograph the relationship between technologies and the people using them—and in between themselves.[10] And indeed, how can we overlook the polity-building processes implied in Google's infrastructure design—its

Figure 3.1. Douglas County, Georgia, data center. Shown here are colorful pipes distributing water for cooling the facility, and Google's G-bike, the "vehicle of choice" for transportation in and around the data centers.

Figure 3.2. Bahnhof data center, Stockholm, Sweden.

lively colored pipes, well-organized lines of glowing server racks in shades of blue and green, and brightly illuminated architectural spaces—as compared with Bahnhof's underground Cold War bunker setting and historical engine for backup power?

Google's data centers literally span the globe, and their images imply a seamless, universal connection, a benevolent global reach, and even a no-impact environmental presence with a corporate-designed bicycle featured in one shot as the "transportation of choice" around the data center. Bahnhof's website, on the other hand, advertises heavily protected security in three separate data centers (one of which is a nuclear bunker) with "TOP SECRET" stamps across the homepage. Further, there are proud proclamations and lengthy explanations about the company's valuing the right to freedom of speech and being the host for Wikileaks, along with a link to the Ebay auction for the Wikileaks data server.[11] Indeed, the images of Bahnhof's data centers speak to us about the ways that Europe's "oldest and strongest legislations for freedom of speech and freedom of information" have been built into the very facilities servicing access to data.[12] In short, such images tell us about affordances and constraints turned into pipes and cables, about in-built political values and the ways the engineering of artifacts come close to engineering via law, rhetoric, and commerce. And the images also testify to the constant struggles over standards and policies intrinsic to the network economy.[13]

Or so we may think. For what is most striking about these images is, of course, precisely that which we do not see. Google's and Bahnhof's images gesture toward the notion of transparency, all while working to conceal or obscure less picturesque dimensions of cloud infrastructure. We learn nothing, in Google's case, about its mechanical, electronic, or technical infrastructure design, energy use, or network infrastructure; in fact, Google is notoriously secretive about the technical details of its servers and networking capabilities in the interest of security as well as competitive strategy.[14] Nor do Bahnhof's photos tell us anything about how much this "free speech" Internet service provider's business actually is built on unauthorized traffic—in Sweden, piracy has been key to the media and IT industries' development, selling conduits and connectivity.[15] Hence, a third and final step is required: we need to acknowledge that many of the operations, standards, and devices we are trying to describe when analyzing digital infrastructure will remain hidden, locked away, or, in engineering terms, "blackboxed." As Bruno Latour has pointed out, the mediating role of techniques is notoriously difficult to measure, at least as long as the machines run smoothly; the more technology succeeds, the more opaque it becomes.[16] Although Google and Bahnhof provide branded services and platforms, and thus are readily apparent to their users, their infrastructures remain blackboxed. Data centers are information infrastructures hiding in plain sight.[17]

This chapter discusses data centers as the material dimension of "the cloud" and as a critical element of digital media infrastructures. To render cloud computing truly visible, we need to understand the material support systems for data storage and data transmission, or the "stuff you can kick," as described by Lisa Parks—the bricks and mortar, physical networks of digital media distribution.[18]

Additionally, we also need to "see" the standards and protocols, affordances and constraints built into these networks. While distribution infrastructures always have been designed to be transparent,[19] transparency as immaterialized in "the cloud" has turned into an all-purpose political metaphor for the fact that we are storing our data (or our company's data) on someone else's servers in an undisclosed location that we will never be able to see. In following media archaeology's "non-representational take on politics," its interest in the "non-sense of something that cannot be exchanged for meaning,"[20] we are turning to what Susan Leigh Star has referred to as the "boring backstage elements"[21] of online delivery, or, in the case of data centers, where "the cloud" touches the ground. Connecting the metaphor and imagery of the cloud to data centers and Internet topology, we aim to discern structures of power through technological and industrial analysis.[22]

The Technopolitics of Hypervisibility

Keeping noves up in the cloud...

Data centers are the heart of "the cloud" and much of its physical infrastructure. They *are* the physical presence of this imaginary space, and yet they strive to remain invisible in many ways. They maintain a high degree of secrecy, allowing very few visitors from the outside in, and keeping their locations, operating procedures, or devices largely out of the press as a matter of security—and competition in the market. In fact, the refusal to discuss where they are located, how many there are, and other details about how and how much data is processed in these centers has led some in the industry to liken the culture of confidentiality surrounding server farms to the ethos of *Fight Club* ("The first rule of data centers is don't talk about the data centers").[23]

One notable exception to this protective veil of secrecy occurred with Google's 2012 public relations push to promote their data centers as visible, accessible, and environmentally friendly. The images of technology on the site devoted to "revealing" their data centers offer colorful shots of computers, wires, routers, switches, pipes, and hard drives that arguably render this infrastructure much less visible when decontextualized. Indeed, it almost appears as abstract art; there is no trace of any relationship between these technological components and the processing, storing, cooling, or distributing trillions of gigabytes (now known as zettabytes) of data—or the attendant environmental implications (see figure 3.3).

The structures where this all takes place have also been hyperstylized to showcase the natural environment and seemingly make the visual argument that the landscape is even *more* beautiful *because of* the giant data center in the picture. There are portraits of lush wildflowers, mist rising above the Columbia

Figure 3.3. Ethernet switches in Google's Berkeley County, South Carolina center.

River gorge, and even deer grazing outside a data center, oblivious to the hulking steel building in their midst (see figures 3.4 and 3.5).

The main foci of the images are the expanse of sky and land surrounding the buildings. In effect, the data centers are visible but rendered practically inconsequential by the surrounding spectacle of natural vistas and wide-open spaces. Bahnhof, on the other hand, is literally embedded in the natural environment. The camouflage of the Swedish data center projects a sense of safety and security by virtue of its carefully constructed invisibility (see figure 3.6).

In many ways, these representational strategies employed by Google and Bahnhof are emblematic of the argument Parks makes in her work on "antenna trees" and the politics of infrastructure visibility: "By disguising infrastructure as part of the natural environment," she writes, "concealment strategies keep citizens naive and uninformed about the network technologies they subsidize and use each day."[24] These traditions of concealment and disguise also render data centers, and digital media infrastructure generally, notoriously difficult to research by applying the toolbox of traditional media industry analysis. Two of

Figure 3.4. The Dalles, Oregon.

Figure 3.5. Council Bluffs, Iowa data center with deer in the foreground.

Figure 3.6. Bahnhof data center, Stockholm, Sweden.

the classical questions of mass communication research—"Which industry?" and "Whose industry?"—seem insufficient when applied to media today.[25]

Digital media infrastructure makes for a case in point. It is difficult to identify clear-cut boundaries between public and private interests in media infrastructures, let alone in between the various businesses providing us with access to media content; nor can we assume that "the industry" follows only one-dimensional strategic goals such as profit maximization. For instance, what we perceive as the quality and service of streamed entertainment is the effect of a complex and ever-changing web of relations that exists at global and local, technical and social, material and experiential levels, involving content as much as content aggregators, services as much as service providers, transport network operators as much as a mushrooming consumer media ecology. This is not anybody's industry in particular; its emergence and change can hardly be pictured in terms of one institution striving for market power. While traditional issues such as concentration of ownership, subsidies and tax breaks, operating efficiencies, and industry resources may remain useful categories for political economic analysis akin to what they were during the first wave of media mergers, today's structural convergence (and functional heterogeneity) of media in a global market demands a more case-based rather than one-size-fits-all approach.

Media infrastructure industries are analytically distinct from traditional media industries as they involve different actors and practices, standards and norms, expectations and tensions, but they are also deeply embedded in our historically grown media cultures. It is thus hardly surprising that some of the most hotly debated questions about digital media infrastructure today concern traditional values about media industry performance based on the understanding of media as a public good, and of media industries as being unlike all other industries.[26] We also expect digitally distributed media not to waste resources, to facilitate free speech and public order, to protect cultural diversity, and to be equitably accessible.[27] We still understand media to be socially more valuable than just household appliances or "toasters with pictures," as former FCC chairman Mark Fowler once controversially put it,[28] while media technologies today indeed mostly come as just that—as cheap, scale-produced hardware add-ons. While we somehow seem to have approved that all other industries produce not only positive but also negative externalities—that is, negative spill-over effects on third parties not involved in the industry's respective market—it appears culturally more challenging to accept the constant overflow caused by industries supplying our alleged demand for what Lev Manovich calls "the stage of More Media."[29]

Thus, while almost anyone in Western economies happily subscribes to the no-cost, cover-it-all promise of a search engine like Google, or to the pleasure of clicking Facebook's like button at anytime anywhere, each of these activities of course comes with consequences for the public-good idea of digital media infrastructure as being shared and sustainable; they are accompanied by rising energy demands, the generation of saleable secondary data, and the like.[30] This results in a situation of policy overlay or "regulatory hangover," where media infrastructures and technologies are framed and identified through an outdated system of rigid, dialectically opposed values (commercial vs. public, open vs. closed, monopolistic vs. competitive, free vs. subscription, formal vs. informal, and so on),[31] while our actual practices and expectations are far more expansive and play havoc with such beliefs. These longstanding and traditional frameworks for evaluating power in media industries grow increasingly limited as communication and information technologies continue to converge. Sandra Braman has explored how this consequent blending of communication styles, media, functions, and industries "disrupts habits of policy analysis," and ultimately our regulatory tools fall short of what is required to effectively maintain current policy goals. As Braman explains, this gap widens as we look at the greater landscape of policy terrain. "The distinction between public and private communicative contexts has become one of choice and will, rather than ownership, control and history of use. And we have come to understand that both non-political content and the infrastructure that carries it can have structural, or constitutive, impact."[32]

Hence, in order to understand how control is exerted through media infrastructure, it's rather naïve to simply ask who owns it.[33] It is similarly limited to assume that a society could actually opt out of globalization processes, choose between a more or less desirable market structure for its media, or push back negative externalities and just enjoy the nice ones. Yet all these reservations do not prevent us from knowing about, and intervening in, the very process through which digital media infrastructures emerge. Our premise is that infrastructures are always relational; they concern materialities as much as technologies and organizations, and they emerge for people in practice.[34] In order to understand today's media infrastructure, we need to study how "distribution is distributed": how it configures (legally and otherwise) the global and local, technical and social in response to a problem that needs a fix.

Studying infrastructure means studying infrastructural relations, but at the same time, infrastructure also is more than just pure matter that enables the movement of other matter, or "the thing other things 'run on.'"[35] As Brian Larkin has pointed out, any infrastructure's peculiar ontology lies precisely in the fact that it forms the relation between things while also being a thing in itself—"as things they are present to the senses, yet they are also displaced in the focus on the matter they move around."[36] Infrastructures are like lenticular prints: they always come with a switch effect ("now you see it, now you don't"), not because they would change in themselves, but because they animate our view, make us shift our categories of what they are—image of connective technologies, image of the technological objects being connected. Data centers may be described as information infrastructures hiding in plain sight in that they resemble such flicker pictures, making us want to explore the depths of what appears to be an image sliding behind another one, the spectacular spaces "behind" the cables and plugs. Yet this exploration is not entirely free or unguided; pleasure is en-gineered into the act of looking itself by divesting the object (rows of server racks, rooms full of water pipes, and so on) from its actual use and turning it into an "excessive fantastic object that generates desire and awe in autonomy of its technical function."[37] This is why it would be insufficient to study only a given media infrastructure's topology, the networks of its relations; the politics of media infrastructure is also in its imaginary. It partly rests on what Larkin calls the "poetic mode" of infrastructure—its capacity to turn us on and away from the objects being connected.[38]

The Google and Bahnhof images referred to above strikingly illustrate this second conceptual premise of our chapter. Infrastructural politics is not just about what is deliberately hidden from sight or is invisible; it is equally about the hypervisibility created around *some* of an infrastructure's component parts, all while most of the relations it engenders and the rationality embodied in its overall system sink deeply in obscurity. If computing has become the privileged technology of our age,[39] then our age is marked by this technology's materiality (silicon, copper, plastics, and the like) as much as by a political form (liberalism) that attempts to organize users and territories through domains that seem far removed from politics. Media infrastructures are indicative of such a mode of governing that disavows itself while at the same time constantly overexposing its material designs in order to represent, for all those who want to see, how modern our possible futures and futures present have become. It is these "politics of 'as if' "[40] that are so overtly discernible in Google's or Bahnhof's intensely stylized images of denuded technologies. In this sense, data centers can be described as persuasive designs: as artifacts that aim to steer user behavior and attitudes in an intended direction while constraining others.[41] It is for these reasons that we also direct our analysis toward the very practices of conceptualizing digital media infrastructure, both in terms of imagery and topology, rather than simply looking at its social, ecological, or economic effects.

Cloud Imaginaries and Energy Requirements

The data that is processed and stored in "the cloud" is vital to the constant flow of news, information, software, and entertainment that populates the digital media landscape. The data *about* this data has become similarly important to defining "the cloud" for the popular imaginary; as the amount of bits being utilized defies comprehension, comparisons to football fields, metaphors about cities, even representations in the form of data scaling Mt. Everest have been drawn in order to make this data and its environment "visible" or understandable in some way (see figure 3.7).

The amount of data that is estimated to be currently stored in the cloud is more than one billion gigabytes; it is also, as one industry report has characterized it, the same as 67 million iPhones worth of data.[42] These and other comparisons give contours (albeit often absurd ones) to the remote storage

Figure 3.7. Image from *State of the Data Center, 2011* infographic. Emerson Network Power, http://www.emersonnetworkpower.com/en/US/About/NewsRoom/Pages/2011DataCenterState.aspx.

capabilities and infrastructure known as "the cloud," which, other than the aforementioned representations of data centers and their decontextualized technologies, remains largely immaterial, dimensionless, and almost impossible to even imagine. Such metaphors also serve to "contain the messy reality" of infrastructure, as described by Star and Lampland.[43] Yet despite these creative numerical valuations, Fuller and Goffey have articulately observed in their analysis of infrastructure's abstractions that "empirical states of fact obtrude but tangentially on the marketing of hyperbole."[44] Indeed, the more sober "facts" about the infrastructure of the cloud rarely collide with the PR-fueled sensational dramatizations and depictions most commonly circulated. Despite the corporate promotion of "cloud computing" and "cloud storage" as abstract, celestial panaceas for managing digital content, there are still considerable concrete, earthbound challenges for this cloud infrastructure as the demand for access to offsite data continues to explode.

The ways that cloud infrastructure is regulated present one significant challenge. Currently, it is almost a legal impossibility to discern, for example, the jurisdiction and often the sovereignty of the data that is processed, stored, circulated, and transmitted by the millions of data centers all over the globe.[45] It is also extremely difficult to regulate various players in the distribution chain of data from storage to consumer, including Content Delivery Networks (CDNs) and Internet Service Providers (ISPs) that have "peering" agreements. Various interconnection points along the distribution chain are notoriously opaque in their reporting of costs, speed, and connection quality. Regulators also lack the metrics and tools necessary to effectively monitor any anti-competitive or anti-consumer behavior in this industry because of the lack of transparency on the part of the companies involved.[46] This arena would benefit from some genuine visibility, as it is often marginalized in the landscapes of infrastructural concerns.

The amount of energy required to power and cool data centers remains chief among those concerns. These facilities are one of the fastest growing consumers of energy, and they are expanding rapidly. In fact, Google's investment alone during 2013 on expanding their centers represents the largest investment in data center infrastructure in the history of the Internet.[47] The resulting energy needs of "the cloud" are indeed astronomical: a single data center can require more power than a medium-size town.[48] According to a recent Greenpeace report examining the energy consumption of data centers and the various components of "cloud power," if the cloud were a country, it would have the fifth largest electricity demand in the world.[49] It has also been estimated that data centers can waste 90 percent of the power that they pull off the grid, and their carbon footprint will likely surpass that of air travel by 2020.[50] The definition of "wasting" power has been debated in this context—a recent *New York Times* investigation found that, on average, data centers were using only 6 percent to 12 percent of the electricity powering their servers to perform computations. The rest was essentially used to keep servers idling and ready in case of a surge in activity that could slow or crash their operations.[51] However, the reserve power is an insurance policy against disaster (in other words, an outage that cuts off all access to the cloud services that largely support the global economy). The value in having this insurance built into the design of data centers is apparently worth the cost to those who own them, revealing much about the logics of cloud infrastructure, which—much like nuclear power plants—are rooted in excess, redundancy, and contingency, governed by the looming specter of worst-case scenarios.

Thanks to these requirements, the proximity to affordable electricity and energy sources are a paramount consideration when determining where to build and locate data centers. The average temperature and climate are also increasingly being factored in to such decisions as more companies try to take advantage of "free cooling" or the use of outside air instead of energy-intensive air conditioning to cool the massive racks of computer servers and prevent them from overheating (which essentially causes the cloud to "disappear"). Lower temperatures outside present significant cost savings inside, as at least half of a data center's energy footprint has historically come from the energy required to keep the servers cool.[52] As a result, there is a growing interdependency between the developing topography of cloud infrastructure and energy politics. Google is at the forefront of this complex relationship, as the company uses roughly one million servers in what has been estimated to be dozens of data centers.[53] Their data center in The Dalles, Oregon, sits on the Columbia River and uses renewable hydropower to run the center. Google is also becoming a growing power "broker"—investing more than $1 billion in clean-power projects (solar plants, wind farms) in order to buy and sell clean electricity and reduce its carbon footprint. Ultimately, the goal is to send more clean power into the local grids near its data centers, "greening" the cloud infrastructure.[54] In the meantime, the company has taken their role as infrastructure provider to new heights, adding *literal* power to the array of global platforms, services, and data centers they provide in order to keep the cloud functional, on their own terms.

The North Carolina "data center corridor," which runs through about seven rural western counties between Charlotte and Asheville, is another case in point highlighting the evolving relationship between infrastructure and energy politics. With major sites owned by Google, Apple, Facebook, Disney, and AT&T, among others, it has emerged as a major hub for cloud infrastructure.[55]

In addition to the tax breaks offered by the economically depressed state, there is an abundance of low-cost power, water for cooling, and a climate that allows for "free cooling" most of the time. However, North Carolina has one of the "dirtiest" electrical grids in the country: it only gets 4 percent of its electricity from renewable sources such as solar, wind, or water; coal and nuclear provide 61 percent and 31 percent of the state's power, respectively. Data centers as an industry have become increasingly targeted by environmental activists for their enormous consumption of (nonrenewable) energy, and as a result, there has been a marked attempt by major cloud-computing companies such as Facebook, Google, and Amazon to promote their cloud infrastructure as embracing clean, green power.

Facebook's newest data center in Luleå, Sweden, is powered entirely by hydroelectric energy. The company has also added detailed pages on their carbon and energy impact, as well as real-time graphic representations of its power and water usage for two data centers in Oregon and North Carolina. The "dashboards" monitor and visualize the centers' Power Usage Effectiveness (PUE) and its Water Usage Effectiveness on dedicated Facebook pages.[56] Apple has gone beyond visualizing their energy usage to pioneering efforts to engineer their own clean energy for their data centers in North Carolina and beyond. In 2012 the company built the world's largest privately owned solar-panel farm to power their Maiden, North Carolina, data center, and they are currently working on another for their facility in Reno, Nevada. Apple's stated goal is to use 100 percent renewable energy at all of their data centers, and by the end of 2012 they were 75 percent there.[57] According to Google's website, roughly one-third of the power the company uses to power its data centers is clean power.[58]

In addition to these digital media industry giants taking on power generation, they are also privatizing the infrastructure for data centers, even those that serve the public sector. Amazon Web Services hosts cloud services for the CIA, the Department of Defense, and the Federal Reserve, to name a few major government clients. Infrastructure so critical to the functioning of our society being privatized and consolidated in the hands of a few major providers has serious potential to end up like the market for ISPs: highly concentrated, consolidated, largely unresponsive to consumer demand or regulators, and operating well outside the parameters of what could ever be labeled "in the public interest." Unmitigated concentration of course also brings with it severe global economic, legal, and political consequences for the free flow of data around the world. Additionally, it begins to invite more centralization of infrastructural authority, which is a troubling move in the direction away from the original end-to-end architectural principle of the Internet.

Mapping the Cloud

3 *layer* !   TM3

Speaking of digital media infrastructure involves imaginaries as much as topologies. Put differently, and by taking up a distinction introduced by media
archeologist Wolfgang Ernst, it involves speaking of signs as much as of signals:
an assessment both of our networks' semantic surfaces and of the data traffic
itself.[59] Concerns about the sustainability of infrastructure are thus reflected
in the overly stylized representations of data centers as much as they are embodied in the Internet's architecture. While the "data about the data" and the
visibility created around newly built data centers are meant to mark a turn away
from the old days of resource-inefficient corporate client-server computing and
toward the net as global public utility,[60] data traffic itself tells different stories.
As a complex engineered system, the Internet includes material, technological, and entrepreneurial arrangements through which telecommunications and
ISPs manage flows of traffic.[61] Turning to the Internet's architecture, we have to
differentiate between three different levels in order to identify how and where
this flow of data gains political (and environmental) implications. The most
obvious layer of Internet architecture consists of overlay networks such as the
World Wide Web, email, or peer-to-peer. Beyond that is the interdomain level;
the Internet is made up of tens of thousands of loosely connected networks
called Autonomous Systems (AS) employing different business models and
profiles (so-called Tier 1 providers, retail services, business services, network
access, Web hosting, and the like). The third layer of architecture is the Internet's physically meaningful topology, that is, the way it builds connectivity at
the router level.[62]

Since it is difficult to assess the relational dimension of digital media infrastructure for a large and diverse country like the United States, we instead turn to
a "small world"[63] like Sweden to elaborate on this issue. Sweden suggests itself
as a case in point not only because of its media ecology's limited size but also
because of the country's above-standard broadband penetration and the fact
that streaming video currently dominates data traffic.[64] Apart from being integrated into services such as Facebook or Google, video can be accessed through
on-demand platforms, and its ubiquity accounts for a major change in digital
media infrastructure when it comes to the first or "user plane" of the Internet.
Sweden is indicative of a global overprovision of over-the-top (OTT) video on-
demand services, offering non-authorized but culturally accepted streaming
(for example, sweafilmer.com) and downloading (The Pirate Bay) platforms
alongside digital public service broadcasting (SVT Play) and advertising-based,
transactional, or subscription streaming services (Netflix, Viaplay, Voddler,

iTunes, and the like). In addition, telco operators such as ComHem or Telia-Sonera provide various Internet protocol television (IPTV) options. This leads to an overprovision not only in terms of accessing services but also, and more important in this context, in terms of the data that is made accessible.

All of the (legal) video platforms have to make content licensing deals with the very same content providers for the very same titles, which then are encoded, stored, and delivered to customers in as many as thirty-two versions per title (as in the case of Netflix), reflecting requests on varying encoding rates, device types (smartphones, tablets, and so on) and Digital Rights Management (DRM) schemes, the latter depending on territorial licensing agreements. A company like Netflix, for instance, which has gained a strong foothold in Sweden, streams one billion hours of content per month to more than 37 million subscribers globally that view it on hundreds of different device types.[65] Given the competition for attention and the streaming costs involved, streaming service providers in Sweden operate with very low to non-existent margins, while a large share of the data they traffic is *redundant*. Traffic is redundant because of the above-described overprovision of identical titles in numerous shapes and by various providers, but also because redundancy is deliberately created by providers in order to ensure the quality of experience in watching any one of their titles. While part of this redundant traffic is necessary to manage varying bandwidths, streaming delay, packet loss, or server failures,[66] a large part of it is indeed unnecessary, wasting network bandwidth and over-utilizing server-side resources.[67] In short, seen from its user end, the "cloud" looks like a pipeline plugged with often inessential and even progressively devalued data—or, in Marc Andrejevic's words, like "an environment of data glut."[68]

While there is limited use in a "close reading" of the router-level or "data plane" of digital media infrastructure in the context of this chapter—that is, of the physical nodes and connections between which data is forwarded based on trafficking policies—a closer look at the secondary or inter-domain level of infrastructure is instructive for assessing the implications of what we have described above. For it is this inter-domain level or "control plane" that configures organizational routing policy—the policies based on what data are sent through the pipes. Data trafficking policy is configured on the inter-domain level either through customer-provider or peering links. To use Sweden as an example, a so-called Tier 1 network provider like TeliaSonera—one of the largest ISPs globally in terms of traffic volume, routes, and autonomous systems inside the network—can provide Internet access at monetary costs similar to a smaller network such as CDNetworks, a content delivery network (CDN) designed to improve the quality of streaming online video. This is called a customer-provider link. Peering links, on the other hand, are bilateral agreements between two AS networks to exchange certain types of traffic free of charge.[69] For instance, in the Swedish case Netflix rents rackspace at neutral local data centers, connecting those servers via so-called IXPs (Internet exchange points, such as Netnod) and employing peering agreements to smaller Swedish broadband or mobile network operators such as ComHem, Bredbandsbolaget, and the like. This way, more than 80 percent of Netflix's data traffic is served from the local Internet service provider's data center, saving the company transit, transport, and other upstream scaling costs.[70] Customer-provider and peering links thus have inherently different business models and trafficking policies.

Even such a cursory description of one smaller European country's infrastructure for video streaming reveals that one indeed needs a relational perspective in order to understand what infrastructure is about. First and most obviously, data centers are merely one element among many; they do not form the one central node in the network from which to unravel digital distribution's mysteries. Second, common concerns about ownership or cultural hegemony obscure rather than enlighten what is critical about digital media infrastructure. Instead of speculating about "the control from one single country over most of the Internet services," as former Pirate Bay spokesman and Internet activist Peter Sunde recently did,[71] we might study the bilateral agreements between foreign and home-grown Internet services, or the way a platform like Netflix is both culturally and technologically *embedded* in Sweden. For why would Swedish ISPs be interested in peering agreements with Netflix? Because Netflix helps them "push the pipe," and, perhaps even more important, because it creates "added value" to their broadband services. When it comes to the political economy of digital infrastructure, we need to look at its specific topology. In order to find out what the policies of streaming video are about, we have to ask: What need is this infrastructure addressing? How does it engineer a solution, and to which problem?

While video streaming technologies are widely marketed as enhancing consumer control over the entertainment-viewing experience ("anytime, anywhere"), their purpose, as it becomes observable on infrastructural level, is primarily to enhance entrepreneurial control over content. Streaming is a technology that allows content providers to "keep" the file rather than distributing it for permanent storage on consumer devices. Control over content is an issue when major autonomous systems such as TeliaSonera push on the digital market themselves by offering content delivery through IPTV and set-top boxes.[72] Control over content pertains to attempts by platforms such as Netflix or Voddler

,ain independence from content delivery networks and telco companies by
.rpose-building their own delivery architectures.[73] Yet this striving for control over information is hardly new, nor is it a strategy solely exerted by private
companies at the expense of public utility. In fact, the most important Scan-
dinavian player on inter-domain level, TeliaSonera, is largely state owned. As
James Beniger documented decades ago in his monumental study, *The Control
Revolution* (1986), there is more continuity than cleavage in the relationship of
today's information society to the past.

* * *

While data centers (and their public profiles) have been rapidly expanding,
the actual infrastructure for media's future still remains woefully insufficient.
It is proliferating, but not as fast as the data it is designed to contain, process,
and distribute. This problem has been characterized by experts as a "race between our ability to create data and our ability to store and manage data."[74]
This race will be one of the true "technological dramas" that will be playing out
in the coming years, as our global media culture is increasingly dependent on
streaming, remote storage, and mobile access. To understand and explain the
many consequences—sociocultural, economic, political, regulatory, and otherwise—of this growing infrastructure gap will require analyses and scholarship
that engages with more than the material dimensions of infrastructure; indeed,
the politics of representation, technology policies, industrial practices, and even
the imaginary, abstract constructions of technologized spaces will all be a part
of bridging—and visualizing—this gap moving forward.

Notes

The authors wish to thank Kurt-Erik Lindquist (Netnod) and Chris Baumann for
helpful suggestions, and the anonymous reviewers for critical comments on an earlier draft on this paper.

1. Google website, "Data Centers," available at http://www.google.com/about/
datacenters (accessed November 27, 2013).

2. See http://www.apple.com/environment/renewable-energy (accessed November 27, 2013).

3. See https://www.facebook.com/PrinevilleDataCenter for one example, or http://
www.globalfoundationservices.com/ for Microsoft (accessed November 27, 2013).

4. "Effectively Minimizing Redundant Internet Streaming Traffic to iOS Devices,"
available at http://www.bahnhof.net/gallery (accessed November 27, 2013).

5. See, for instance, Douglas Alger's book *The Art of the Data Center* (New Jersey:
Pearson Education, 2013); the documentaries *Bundled, Buried and Behind Closed Doors*
(Mendelsohn, 2011) and *The Bucket Brigade* (Mick, 2012); or Andrew Blum's *Tubes: A
Journey to the Center of the Internet* (New York: HarperCollins, 2012).

6. Peter Jakobsson and Fredrik Stiernstedt, "Time, Space and Clouds of Information: Data Center Discourse and the Meaning of Durability," in *Cultural Technologies: The Shaping of Culture in Media and Society*, ed. Göran Bolin (New York: Routledge, 2012), 103–18. See also Peter Jakobsson and Fredrik Stiernstedt, "Googleplex and Informational Culture" in *Media Houses: Architecture, Media, and the Production of Centrality*, ed. Staffan Ericsson and Kristina Riegert (New York: Peter Lang, 2014), 112–32.

7. Wakefield Research, "Citrix Cloud Survey," August 2012, 1, available at http://www.citrix.com/content/dam/citrix/en_us/documents/go/Citrix-Cloud-Survey-Guide.pdf (accessed November 27, 2013).

8. Culture Digitally Blog, post by Casey O'Donnell at http://www.marco.org/2013/07/03/lockdown (accessed November 27, 2013).

9. Bryan Pfaffenberger, "Technological Dramas," *Science, Technology and Human Values* 17, no. 3 (Summer, 1992): 285; compare with Finn Brunton, *Spam: A Shadow History of the Internet* (Cambridge, Mass.: MIT Press, 2013), xvi.

10. Josh Braun, "Going over the Top: Online Television Distribution as Sociotechnical System," *Communication, Culture and Critique* 6 (2013): 432–58.

11. Bahnhof website, http://www.bahnhof.net/wikileaks (accessed November 27, 2013).

12. Bahnhof website, http://www.bahnhof.net/basedinsweden (accessed November 27, 2013).

13. Jan van Dijk, *The Network Society* (Thousand Oaks, Calif.: Sage, 2012), 79.

14. See http://www.wired.com/wiredenterprise/2012/03/google-miner-helmet (accessed November 27, 2013). Google's public relations team denied all requests for interviews about their data centers for this chapter.

15. Jonathan Sterne, *mp3: The Meaning of a Format* (Durham, N.C.: Duke University Press, 2012), 209, 219.

16. Bruno Latour, "On Technical Mediation," *Common Knowledge* 3, no. 2 (1994): 29–64; see also Bruno Latour, *Pandora's Hope: Essays on the Reality of Science Studies* (Cambridge, Mass.: Harvard University Press, 1999).

17. Compare with Braun's notion of "transparent intermediaries" (2013), which seems less useful here, however, as it relates transparency primarily to white label platforms/services.

18. Lisa Parks, " 'Stuff You Can Kick': Toward a Theory of Media Infrastructures," in *Humanities and the Digital*, ed. David Theo Goldberg and Patrik Svensson (Cambridge, Mass.: MIT Press, 2015), 355–73.

19. Lisa Parks, "Where the Cable Ends: Television in Fringe Areas," in *Cable Visions: Television Beyond Broadcasting,* eds. Sarah Banet-Weiser, Cynthia Chris, and Anthony Freitas (New York: New York University Press, 2006), 124.

20. Jussi Parikka, "Critically Engineered Wireless Politics," *Culture Machine* 14 (2013): 1–26.

21. Susan Leigh Star and Martha Lampland, eds., *Standards and Their Stories: How Quantifying, Classifying and Formalizing Practices Shape Everyday Life* (Ithaca, N.Y.: Cornell University Press, 2009).

22. Parikka, "Critically Engineered Wireless Politics," 164.

23. Quoted in Blum, *Tubes*, 239.

24. Lisa Parks, "Around the Antenna Tree: The Politics of Infrastructural Visibility," *Flow*, March 6, 2009, available at http://flowtv.org/2009/03/around-the-antenna-tree-the-politics-of-infrastructural-visibilitylisa-parks-uc-santa-barbara (accessed November 27, 2013).

25. See, for instance, Horace Newcomb, "Toward Synthetic Media Industry Research" in *Media Industries: History, Theory, and Method*, ed. Alisa Perren and Jennifer Holt (Chichester: Wiley-Blackwell, 2009), 264–71.

26. For more on these fundamental questions related to media markets, see David Hesmondalgh, *The Cultural Industries* (London: Sage, 2013); Jeff Chester, *Digital Destiny* (New York: New Press, 2007); Des Freedman, "Outsourcing Internet Regulation," and "Web 2.0 and the Death of the Blockbuster Economy," in *Misunderstanding the Internet*, ed. James Curran, Natalie Fenton, and Des Freedman (New York: Routledge, 2012), 69–94, 95–120; Terry Flew, *The Creative Industries* (Los Angeles: Sage, 2012); Sean Cubitt, "Distribution and Media Flows," *Cultural Politics* 1, no. 2 (2005), 193–214; Arthur de Vany, *Hollywood Economics: How Extreme Uncertainty Shapes the Film Industry* (London: Routledge, 2004); Harold Vogel, *Entertainment Industry Economics: A Guide for Financial Analysis* (Cambridge: Cambridge University Press, 2010).

27. Benjamin Compaine and Douglas Gomery, *Who Owns the Media? Competition and Concentration in the Mass Media Industry* (London: Erlbaum, 2000), 523. See also Ithiel de Sola Pool, *Technologies of Freedom* (Cambridge, Mass.: Harvard University Press, 1983); Robert McChesney, *Digital Disconnect* (New York: New Press, 2013); Susan Crawford, *Captive Audience* (New Haven, Conn.: Yale University Press, 2013).

28. Mark Fowler, quoted in David Croteau and William Hoynes, *The Business of Media: Corporate Media and the Public Interest* (New York: Pine Forge, 2006), 25.

29. Lev Manovich, "Cultural Analytics: Visualing Cultural Patterns in the Era of "More Media," in *Domus* (Milan, 2009); for the notion of overflow, see Michel Callon, "An Essay on Framing and Overflowing: Economic Externalities Revisited by Sociology" in *On Markets*, ed. John Law (Oxford / Keele: Blackwell / Sociological Review, 1998), 244–69.

30. Richard Maxwell and Toby Miller, *Greening the Media* (Oxford: Oxford University Press, 2012); Nadia Bozak, *The Cinematic Footprint: Lights, Camera, Natural Resources* (New Brunswick, N.J.: Rutgers University Press, 2012); Sean Cubitt, Robert Hassan, and Ingrid Volkmer, "Does Cloud Computing Have a Silver Lining?" *Media, Culture and Society* 33, no. 1 (2011): 149–58 and elsewhere.

31. See Vogel, *Entertainment Industry Economics*, 528, for a list of conventional policy issues; compare with Croteau and Hoynes, *Business of Media*, 25.

32. Sandra Braman, "Where Has Media Policy Gone? Defining the Field in the Twenty-First Century," *Communicaiton Law and Policy* 9, no. 2 (Spring 2004): 156, 159.

33. See, for instance, Jack Goldsmith and Tim Wu, *Who Controls the Internet? Illusions of a Borderless World* (Oxford: Oxford University Press, 2006).

34. Geoffrey C. Bowker, Karen Baker, Florence Millerand, and David Ribes, "Toward Information Infrastructure Studies: Ways of Knowing in a Networked Environment," in *The International Handbook of Internet Research*, ed. Jeremy Hunsinger, Lisbeth Klastrup, and Matthew Allen (Heidelberg: Springer, 2010), 97–117.

35. Star and Lampland, *Standards and Their Stories*, 17.

36. Brian Larkin, "The Politics and Poetics of Infrastructure," *Annual Review of Anthropology* 42 (2013): 329.

37. Ibid., 333.

38. Ibid., 335.

39. Compare with Lewis Mumford, *Technics and Civilization* (Chicago: Chicago University Press, 2010).

40. Larkin, *Politics and Poetics*, 335.

41. Persuasive design has mostly been propelled by computer technologies engineering user behavior, such as prescriptive social software like Brightkite or Loopt. See Alice E. Marwick, "Foursquare, Locative Media, and Prescriptive Social Media," 2009, available at http://www.tiara.org (accessed November 27, 2013).

42. Nasuni, "The State of Cloud Storage 2013 Industry Report," available at http://www6.nasuni.com/rs/nasuni/images/2013_Nasuni_CSP_Report.pdf, and "Comparing Cloud Storgage Providers in 2013," infographic, available at http://www.nasuni.com/blog/193-comparing_cloud_storage_providers_in (accessed November 27, 2013).

43. Star and Lampland, *Standards and Their Stories*, 11.

44. Matthew Fuller and Andrew Goffey, *Evil Media* (Cambridge, Mass.: MIT Press, 2012), 58.

45. See Jennifer Holt, "Regulating Connected Viewing," in *Connected Viewing: Selling, Streaming and Sharing Digital Media*, ed. Jennifer Holt and Kevin Sanson (New York: Routledge, 2013), 19–39; and Primavera De Filippi and Smári McCarthy, "Cloud Computing: Centralization and Data Sovereignty," *European Journal for Law and Technology* 3, no. 2 (2012), available at http://ejlt.org//article/view/101/234 (accessed November 27, 2013).

46. See, for example, Stacey Higginbotham, "Video Isn't Breaking the Internet: The Industry Giants Are," *GigaOm*, September 16, 2013, available at http://gigaom.com/2013/09/16/video-isnt-breaking-the-internet-the-industry-giants-are (accessed November 27, 2013).

47. Rich Miller, "Google's Infrastructure Boom Continues: Expansion Ahead in Oregon," *Data Center Knowledge*, April 24, 2013, available at http://www.datacenterknowledge.com/archives/2013/04/24/googles-infrastructure-boom-continues-expansion-ahead-in-oregon (accessed November 27, 2013).

48. Quoted in James Glanz, "Power, Pollution and the Internet," *New York Times*, September 22, 2012, available at http://www.nytimes.com/2012/09/23/technology/data-centers-waste-vast-amounts-of-energy-belying-industry-image.html?smid=tw-share&pagewanted=all&_r=1& (accessed November 27, 2013).

49. Quoted in Gary Cook, "How Clean Is Your Cloud?" *Greenpeace International*, April 2012, 10, available at http://www.greenpeace.org/international/Global/international/publications/climate/2012/iCoal/HowCleanisYourCloud.pdf (accessed November 27, 2013).

50. See Glanz, "Power, Pollution and the Internet"; and Cubbitt, Hassan, and Volkmer, "Silver Lining," 154.

51. Glanz, "Power, Pollution and the Internet."

52. Cook, "How Clean," 23.

53. Google's server farms are all over the United States, as well as in Canada, Germany, The Netherlands, France, the United Kingdom, Italy, Russia, Brazil, Tokyo, Beijing, and Hong Kong. See http://www.datacenterknowledge.com/archives/2012/05/15/google-data-center-faq (accessed September 20, 2014).

54. See Ucilia Wang, "Google the Power Player Invests in Another Giant Solar Farm," *Forbes*, October 10, 2013, available at http://www.forbes.com/sites/uciliawang/2013/10/10/google-the-power-player-invests-in-another-giant-solar-farm (accessed November 27, 2013).

55. See Katie Fehrenbacher, "10 Reasons Apple, Facebook, and Google Chose North Carolina for Their Mega Data Centers," *Giga Om*, July 10, 2012, available at http://gigaom.com/2012/07/10/10-reasons-apple-facebook-google-chose-north-carolina-for-their-mega-data-centers (accessed November 27, 2013).

56. The visualization for Facebook's Prineville, Oregon, center can be found at https://www.facebook.com/PrinevilleDataCenter/app_399244020173259 (accessed November 27, 2013). The Forest City, North Carolina, data center can be found at https://www.facebook.com/ForestCityDataCenter/app_288655784601722 (accessed November 27, 2013). The page detailing the company's carbon and energy impact can be found at https://www.fb-carbon.com/pdf/FB_carbon_enegergy_impact_2012.pdf (accessed November 27, 2013).

57. Apple owns only four data centers at this writing: Maiden, North Carolina; Prineville, Oregon; Newark, California; Reno, Nevada; there are reports of a Hong Kong facility being built. Achieving 100 percent renewable energy is a much smaller, albeit significant, task than it would be for a company like Google. See "Worldwide Renewable Energy Use at Apple" graph, Apple and the Environment website, http://www.apple.com/environment/renewable-energy (accessed November 27, 2013).

58. Google website, "Green," http://www.google.com/green/bigpicture (accessed November 27, 2013).

59. Wolfgang Ernst: "Between Real Time and Memory on Demand: Reflections on/of Television," *South Atlantic Quarterly* 101, no. 3 (Summer 2002): 627–28.

60. One should not forget that today's data centers are related to, if not caused by, another energy problem: the increasing power costs of data centers *within* corporate environment or even industrial sectors, and the overbuilding of IT assets in every industry sector. See Nicholas Carr, *The Big Switch: Rewiring the World, from Edison to Google* (New York: Norton), 56.

61. Paul Dourish and Genevieve Bell, "The Infrastructure of Experience and the Experience of Infrastructure: Meaning and Structure in Everyday Encounters with Space," *Environment and Planning B: Planning and Design* 34 (2007): 416.

62. This definition roughly correlates with a common computer science distinction between data plane and control plane, and we are aware of that difference in our terminology.

63. In the sense of Jeffrey Travers and Stanley Milgram, "An Experimental Study of the Small World Problem," *Sociometry* 32, no. 4 (1969): 425–43.

64. See Patrick Vonderau, "Beyond Piracy: Understanding Digital Markets," in *Connected Viewing: Selling, Streaming, and Sharing Media in the Digital Era*, ed. Jennifer Holt and Kevin Sanson (New York: Routledge, 2013), 99–123. Thanks to Kurtis Lindqvist (Netnod.se, Sweden) for additional information on Sweden's digital infrastructure.

65. Nina Hjorth Bargisen (European network strategy manager for Netflix), presentation given at the Netnod Members Meeting, 2013.

66. Sanjoy Paul, *Digital Video Distribution in Broadband, Television, Mobile and Converged Networks. Trends, Challenges and Solutions* (Chichester: Wiley, 2011), 53.

67. Yao Liu, Fei Li, Lei Guo et. al., "Effectively Minimizing Redundant Internet Streaming Traffic to iOS Devices," IEEE Transactions on Multimedia (2013), available at http://www.cs.binghamton.edu/~yaoliu/publications/infocom13-range.pdf (accessed November 27, 2013).

68. Andrejevic, 41.

69. Compare with Dourish and Bell, "Infrastructure," 416; they define common carrier and bilateral peering as "arrangements through which telecommunications and Internet service providers manage flows of traffic."

70. Bargisen, presentation.

71. Peter Sunde (brokep), "Intertubes," available at http://blog.brokep.com/2013/05/26/intertubes (accessed November 27, 2013).

72. Backed by private equity group EQT, TeliaSonera currently aims to gain control over Swedish telecommunication operators and to establish itself as an infrastructural gatekeeper. See Per Lindvall, "Låt staten ta hand om nätkontrollen," *Svenska Dagbladet*, February 3, 2014.

73. Voddler, for instance, uses its own patented streaming solution VoddlerNet, while Netflix has built up its own content delivery network, OpenConnect.

74. Quoted in Glanz, "Power, Pollution and the Internet."

CHAPTER 4

Deep Time of Media Infrastructure

SHANNON MATTERN

When it first appeared in English usage in the mid-1920s, "infrastructure" referred to roads, tunnels, other public works, and permanent military structures. Google's Ngram viewer, which displays the frequency with which words appear in Google's corpora of books, shows that the term was rather obscure until around 1960—roughly the same time that "media" began to take off and "telecommunications" came into widespread use. Thus it is no coincidence that infrastructure—a word whose Latin roots, denoting any form of substructure, would seem to lend it to liberal use—is commonly associated with modern electronic communications and the trafficking of audiovisual signals.

Yet those trafficked signals long precede the age of telecommunication. And infrastructure itself has a much longer history: it has existed as long as has civilization. In fact, we could say that infrastructures made human settlement possible. I am speaking not only of roads and aqueducts and sewers, the kinds of infrastructures that archaeologists and ancient historians commonly examine. *Media* infrastructures, too, have been integrated into our cities, either by design or by accident, since the days of Eridu and Uruk. Anthropologist Clifford Geertz, urban historian Peter Hall, and archaeologist Paul Wheatley all suggest that the birth of cities is rooted just as much in the need for ceremony and communication as it is in economics, which is the prevailing theory.[1] Thus, early cities had to provide spaces conducive to pageantry and communication. Lewis Mumford, author of two grand histories of urbanity, agrees that "what transform[ed] the passive agricultural regimes of the village into the active in-stitutions of the city" was not merely a growth in size or population density or economy, but an extension of "the area of local intercourse, that engenders the need for combination and co-operation, communication and communion."[2] That "area of local intercourse" is an infrastructure—a structure that undergirds communication and communion.

By rethinking what constitutes a media infrastructure, and by acknowledg-ing its deep history, I hope to provide a useful counterpoint to the other stud-ies in this volume. I want to think beyond telecommunications, beyond the nineteenth century, back beyond those technological systems administered by modern states, governmental agencies, and multinational corporations. Tak-ing inspiration from the field of geology and the work of Siegfried Zielinski, we—media and infrastructure scholars, urban historians, even engineers and urban designers—would do well to look at the *deep time* of media infrastructure.[3] And in this more expansive thinking, I want those of us in media and design studies to consider what we might learn from fields of study and practice that have long been examining infrastructure, but which have had little contact with our field. Archaeology and urban and architectural history in particular have much to offer the study of signal traffic. Of course, media studies has already witnessed the arrival of a subfield called "media archaeology," involving such figures as Zielinski, Friedrich Kittler, Erkki Huhtamo, and Jussi Parikka—and while this work does offer an alternative, nonlinear, materialist means of writ-ing media histories, it regards archaeology metaphorically or methodologically rather than literally. I want instead to consider insights from trowel-wielding archaeologists.

Infrastructure historian Paul Edwards admits that, today, infrastructure "has become a slippery term, often used to mean essentially any important, widely shared, human-constructed resource"; this could include hardware, organiza-tions, "socially communicated background knowledge"—any sociotechnical systems that offer "near-ubiquitous accessibility."[4] Despite, or perhaps *because* of, the flexibility of the term, I think we in media and design studies have much to learn from the way Edwards and other historians and theorists of infrastructure conceive of and work with their object of study. In the next section I examine what archaeologists and urban and architectural historians can tell us about how ancient cities provided infrastructures for vocality—for public address and conversation—and for writing. And in the final section I explore how these other fields' methods, or conceptual units, resonate with the historiographic approaches of media studies and can encourage us to reflect critically on how we construct our media—as well as our urban and architectural—histories.

My goal is to demonstrate both how thinking in terms of infrastructure can enhance existing research within media studies—particularly work on the "media city"—and how thinking in terms of the urban environment can elongate our historical view of media infrastructure and allow us to understand more broadly what constitutes a media infrastructure. What can be gained by looking back to the deep time of media infrastructure and its role in engendering and shaping our cities? First, from the perspective of media scholars, we can appreciate media as potentially embodied on a macro scale, as a force whose modes and ideologies and aesthetics of operation can be spatialized, and materialized, in the landscape. We can read the archaeological record, conduct forensic analyses—or, when we are dealing with a medium like the voice, for which there is no collectable artifact, we can use techniques from archaeoacoustics to "listen" to spaces past. We can dig up the cables, pull out the wires, trace the epigraphy on building facades, analyze the disks—and then observe their layering and interconnection.

And when examining media at the macro scale, we also have to acknowledge that media's history is entwined with that of our cities, their streets and buildings, their political-economic and social networks, and so on. In the process, we come to realize that those cities carry in them the "residue" of all media technologies past—and that, furthermore, these "past" media are not merely artifacts or ruins. Much like Raymond Williams's category of the "residual," they are "formed in the past, but . . . still active in the cultural process, not only and often not at all as an element of the past, but as an effective element of the present."[5] This is why our cities today are not solely virtual but are simultaneously aural, graphic, textual, sonic, visual, and digital. We tend in media studies to write format-specific histories, and to suggest that new technologies supplant the old—but when we look at our media histories through our cities, we observe a layering, or resounding, of media epochs. Such realizations open up new methodological opportunities for studying media.

Second, work on infrastructure has the potential to contribute to urban and architectural history, too. For instance, it is possible to reevaluate theories about the birth of cities, which tend to privilege economic explanations for urbanization, and reinforce the central role played by media and communication in urban history. Furthermore, we can highlight the role of communication in giving *form* to our cities. Prevailing theories suggest that urban form is shaped primarily by topography, transportation, defense, or even cosmological or philosophical views. Yet various means of communication—whether the voice or print or digital technologies—have also shaped cities throughout history.

Deep Time of the Media City

There is a well-established but ever growing area of study within media studies that seems to lend itself to the interdisciplinary study I am proposing here. Scholars focusing on the "media city" have tapped into insights from architectural and urban history and theory in order to think about media in relation to "the urban," yet they have tended to focus their attention on *modern* media—photography, film, television, and the like. There is a plethora of research on architecture and cities in relation to mechanically reproduced still and moving images. For instance, many photographic, architectural, and cultural historians, inspired greatly by Walter Benjamin, have examined the city as a photographic subject, photography's early role in the documentation of urban transformation and as an instigator of social change, and photography's influence on particular modern architectural and urban designers.[6] There is also a tremendous amount of work on the city and film as contemporaneous developments, the representation of the city *in* film (this is the dominant thread by far), and film's influence upon architects and planners, including some investigations of the city as a physical and social infrastructure for the rise of film.[7] In more recent decades, scholars like Lynn Spigel and Anna McCarthy have begun to address the synchronous rise of television and postwar suburbs, the politics of screens in public places, and the impact of networked digital media on urban design and urban experience.[8] There has also been excellent work on the impact of radio and modern sound technologies on architecture, zoning, and urban experience.[9]

Some media-cities research evinces an assumption that the mediation of the city *began* with modern media. Scott McQuire, in *The Media City*, observes that the mediation of urban experience "has been underway at least since the development of technological images in the context of urban 'modernization' in the mid-nineteenth century."[10] Eric Gordon, in *The Urban Spectator*, locates the origin of the media city even later than does McQuire: "from the hand-held camera at the end of the nineteenth century to the mobile phone at the end of the twentieth, the city has always been a mediated construct."[11] I contend that "always" begins well before the late nineteenth century and the era of telecommunications.

Cities have, of course, been *represented* for millennia in maps, paintings, woodcuts, lyric poems, and other media formats. Yet the city as a "mediated construct" certainly encompasses much more than mere portrayals of the city; media technologies—particularly media infrastructures—have been embedded in and informing the morphological evolution of our cities since their coming into being. The "media cities" research very rarely looks at infrastructure.

That has changed a bit during the past two decades, with the rise of digital and locative media and ubiquitous computing, which has inspired scholars, designers, engineers, and artists to turn their attention to the technical networks that make new forms of urban mediation possible. But these scholars and practitioners rarely look back to see the technical networks that have always been there, making cities communicative spaces. There is a tendency to overlook the infrastructures that precede the "cyber" and the electronic, as well as those systems that emerged even before the term "infrastructure" itself.

In the fifteenth century, for example, as architectural historian Mario Carpo has explained, new printing technologies brought with them new infrastructures of publishing and education that dramatically influenced design practices.[12] Publishing centers, with their embedded political-economic, social, and technical infrastructures, arose in cities across Asia and Europe. The emergence of new print forms also influenced how people navigated and comprehended their cities. Even today, metaphors of the book inform how we "write" and "read" our cities. Planners talk of "legible urbanism" and of reading the city as a "text," while designers build augmented reality applications layering text and image atop views of the city, making possible a palimpsestic urban "reading."[13]

The voice, too, has long been built into urban form. Since their very beginnings, cities have been places of public address and conversation, and acoustic considerations have, to some degree, informed design and construction. Yet if we look back to the agora of Athens or the Forum in Rome, we will not find infrastructures in the form of electrical wiring and public address systems and stages with acoustic paneling. Instead, as I argue elsewhere, urban surfaces, volumes, and voids have functioned as sounding boards and resonance chambers for mediation, and as transmission media themselves (much of the following discussion on oral communication draws from that previous publication).[14] Particularly in cases like these, media scholars can benefit from the work of archaeologists by excavating the urban contexts and deep pasts of media infrastructures. For instance, archaeology and its subfield of archaeoacoustics, along with architectural and urban history, can enhance understanding of the ways in which these material spaces have, either by design or by accident (archaeologists and architectural historians disagree on the intentionality of ancient acoustic design), functioned as infrastructures of speech and vocality.

"Never in my opinion," Quintilian writes, "would the founders of cities have induced their unsettled multitudes to form communities had they not moved them by the magic of their eloquence."[15] Aristotle, likewise, prescribed a city that would contain no more people than could hear a herald's voice, and architect Vitruvius tells us in the first century BC of fellow designers who sought to cultivate acoustics that maximized the "clearness and sweetness" of orators' voices.[16] Architectural historian Diane Favro and classicist Christopher Johanson are creating digital models of the Roman Forum to understand how the space accommodated funeral processions, and, in part, how it functioned *acoustically* as a space for speech.[17] We find similar acoustic concerns even earlier, in ancient Greece. Classicist Christopher Johnstone has drawn on archaeological research to explore how the architecture of Athens's agora, and, later, civic buildings like the stoa, law courts, and various auditoria shaped both an orator's delivery and his audience's engagement—and even limited the size of the audience, which might be a governing body or jury.[18] These urban volumes thus undergirded the central modality of communication and therefore became a means of governance and a prime medium for sociality in ancient civilization.

What about a city whose infrastructures were formed millennia later, in a different age of media infrastructure? Consider New York in the mid-nineteenth century, when, as David Henkins writes in *City Reading*, mass-produced print was plastered all over the city in the form of posters, signs, and newspapers. During this period the mechanically reproduced image was gaining popularity and telecommunications were rising.[19] Even then, the city was a place of public address; the "residual" medium of oral communications was still shaping urban morphology. Samuel Ruggles, one of the developers of New York's Union

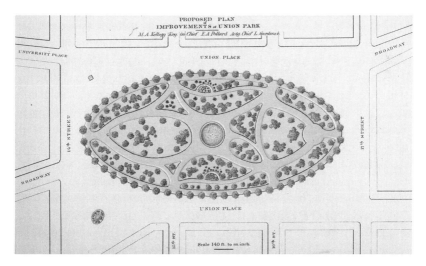

Figure 4.1. Proposed plan improvements of Union Park; by Charles Spangenberg. From New York City Parks Department *Annual Report* (New York, 1871); Mid-Manhattan Library Picture Collection, New York Public Library.

Square, claimed in 1864 that the square was "deliberately designed to support participatory democracy. The triangular parcels of land left over by the imposition of the elliptical park on the grid were expressly made for 'the assemblage of large masses of our citizens in public meetings.' "[20] Through its continual renovation, planners aimed to use the square as an infrastructure to create "active and informed citizens as well as foster social harmony," yet it remained, and *remains*, a site for radical meetings and rallies. Today, Union Square, like many squares and plazas in Athens and Rome and other ancient cities of the world, serves as an urban infrastructure for the integration of a variety of media: locative technologies, text messages, cloth banners, and, still, the bull-horned or naked human voice.[21]

Infrastructures of writing have also long informed how cities took shape. Of course, the first writing surfaces, made of clay and stone, were the same materials used to construct ancient buildings. And often those building facades were the substrates for written texts. The "epigraphic habit" distinguished ancient Greece and Rome. "The Romans seemed to inscribe into everything," according to Johanson. Around the Forum an ancient could find "the written word covering every surface of every major monument."[22] These monuments and building facades were not *designed* to be used as substrates for writing—as an architectural infrastructure for communication—but through the Romans' social practices, "the fabric of the city" ultimately served to record major laws, achievements, legal transactions, and other missives. The city was "informally archiving itself on its skin." Archaeologist Louise Revell acknowledges that such epigraphs constitute a "natural adjunct" to the public architecture on which they were posted or inscribed; the writings played an integral part of political processes and religious services and thus were bound up in the social practice of what it was to be Roman.[23] It is important to note that this "mediation" of Roman identity did not adhere to a single modality; the Forum provided an infrastructure for the public performance or presentation of multiple modes of communication—public address, inscription, sculpture, and other forms of multimedia pageantry. The same can be said of ancient forums adapted for contemporary use, although today's media mix now includes digital technologies among the analog.

The Arabic world has been similarly rich with epigraphy. Art historian Irene Bierman writes of how, in the tenth through the twelfth centuries, the Fatimids of Cairo displayed official writing on the exteriors of minarets and other public structures. Thus, as in Greece and Rome, architecture functioned as an infrastructure for communicating territorial claims and codifying beliefs, and, as Bierman argues, the specific aesthetic properties of those "public texts"— their "color, materiality, and form"—played a key role in how and what they communicated.[24] Art historian Robert Harrist makes similar claims about Chinese *moya*, writings in stone that functioned as "landscape" texts and that, "through their placement in and their interaction with the natural world, both embed historical memory in the topography of China and evoke mythic worlds that transcend the experiences of everyday life."[25]

But writing is not merely inscribed *on* our cities' walls. Lewis Mumford and Harold Innis discuss writing as central to the rise of trade, accountancy, and governance, and thus to the administration of the first cities. Writing is an integral urban political-economic infrastructure.[26] Anthropologist Brinkley Messick argues that we can even find parallels between writing and urban form. He examines the history of Islamic architectural inscriptions and their formal parallels in the very "articulation" of urban space.[27] Messick discusses so-called Arabic "spiral texts," texts in which the writing rotates in a spiral shape, entwining form and content, and he argues that "this poetics of written space then can be extended to general domains of spatial organization: towns, architecture, and the space of the state."[28] He contrasts the "curvilinear urban script" of the Yemeni town of Ibb—which he describes as "a labyrinth of closely packed multistoried houses on narrow and winding alleys and culs-de-sac," with plenty of "residual, irregular spaces"—with the zoned, planned-out newer regions, characterized

Figure 4.2. Spiral urban form in Ta'izz, Yemen. Bezur, Ta'izz, with Aschrifayya Mosque, Wikimedia Commons, CC BY-SA 3.0, http://en.wikipedia.org/wiki/File:Taizz.jpg#file.

by "relatively straight-line, wide thoroughfares with some space left between the buildings."[29] This "new separation and precedence of urban form over urban content," and the parallel evolution in urban form, he argues, "is analogous to the changeover from spiral texts to their straightened successors." Whether we can claim a *causal* relationship is perhaps beside the point; what we see here is a morphological resonance between an integral political-economic and cultural media-infrastructure and the shape of the city itself.

Today, architectural and urban theorists seem ready to posit deterministic formal relationships between digital infrastructures and our new "smart cities" in Songdo, Korea, or Masdar, U.A.E. The builders of these networked developments often design *out* opportunities for unplanned (and un-modern) modes of communication: streets seem intended primarily to shuttle people from one telecommunication station to another, rather than to foster face-to-face interactions; and building facades are constructed of anti-graffiti materials. It seems that, in such places, there is little "residual" media infrastructure to dig into. Yet there has already arisen a huge contingent of critics who argue that such developments, by contradicting millennia of urban design experience, are destined to fail.[30] As Richard Sennett has argued, these over-zoned, over-rationalized cities, devoid of any historical sensibility, defy "the fact that real development in cities is often haphazard, or in between the cracks of what's allowed."[31] "The danger now is that this information-rich city may do nothing to help people think for themselves or communicate well with one another." A media city that makes no provisions for a layering of communicative infrastructures, that wipes away the deep time of urban mediation, is more stupefying than smart, more machine than metropolis.

These examples demonstrate that our media histories are deeply "networked" with our urban and architectural histories (and *futures*) and that, in many cases, these cultural and technological forms are mutually constructed. Thus, particularly in studying the deep time of our media infrastructures, scholars and practitioners in *all* fields need to regard these systems *in relation* to one another. What's more, we need to recognize that the integration of these various structures simultaneously shapes, and is shaped by, the social practices and everyday experiences of the people who live with them.

Methods for Digging into Infrastructure's Deep Time

In this final section, I examine how archaeology's, urban history's, and infrastructure studies' methods and central concepts resonate with the methodological approaches of media studies, and how these "imported" intellectual tools might encourage us to think more critically about how we construct our media histories. I hope also to reinforce suggestions made in the previous section—that thinking in terms of infrastructure can enhance media studies research on the "media city," and thinking in terms of the urban environment can extend the historical scope of media infrastructure and allow a broader understanding of what constitutes an infrastructure. In what follows, I outline six historiographic or methodological lessons that emerge from thinking about the media-city in relation to infrastructure and from thinking about media infrastructure in relation to urban history. By no means are these six lessons, or concepts, mutually exclusive; there is actually a good bit of redundancy, but I think that, in some cases, restating the same principle using different language can only enhance its potential utility.

TECHNO-SOCIO-SPATIO-MATERIAL ENTANGLEMENTS

The deep time of urban mediation is manifested in material strata—in literal *layering*. Henri Lefebvre has argued that urban space is formed by superimposed capital regimes and the infrastructures they create in their own image; the result, he has famously suggested, is not unlike a flaky *mille-feuille* pastry.[32] But the palimpsest is not a mere metaphor. In his excellent study of infrastructure in urban Nigeria, anthropologist Brian Larkin writes that the "physical shape of the city emerges from the layering of . . . infrastructures over time."[33] The nature of that layering, however, is not one of mere supplanting or obsolescence. If we dig down through the strata, we find much more than *ruins* (and this is where, I think, the archaeological metaphor can at times be a bit misleading). Digging into these layers, we often find that, depending on different contextual factors, various infrastructures have distinctive temporalities and evolutionary paths.

As I have argued elsewhere, through "excavation" we can assess the life-spans of various media infrastructures and determine when "old" infrastructures "leak" into new-media landscapes, when media of different epochs are layered palimpsestically, or when new infrastructures "remediate" their predecessors.[34] Geographers Stephen Graham and Simon Marvin write that "because of the costs of developing new telecommunications networks," for instance, "all efforts are made to string optic fibers through water, gas, and sewage ducts; between cities, existing railway, road, and waterway routes are often used."[35] And in the Roman Forum, as Johanson explained, sculpture, architecture, epigraphy, and public address all reinforced one another in the spectacle of the funeral procession and other public pageantry. The same architecture that served as a sounding board for public address also served as a substrate for epigraphy— and today serves as a substrate for graffiti and as a scaffolding for cell phone antennae. The historical media infrastructures on the "lower levels" of our cities are often very much alive in, and continuing to shape, the contemporary city. They are Williams's "residuals." This intermingling of temporalities fits Christopher Witmore's definition of "archaeological time": "the entanglement, the intermingling, the chiasm of pasts and presents."[36]

NETWORKED HISTORIES

Graham and Marvin list some of those intermingling—"superimposed, contested and interconnecting"—infrastructural layers, or what they call "scapes": the "'electropolis' of energy and power," "the 'hydropolis' of water and waste," "the 'cybercity' of electronic communication."[37] But by taking the long view on this intermingling, it is possible to understand these "scapes" as tangled up with one another not only spatially but also temporally. The history of any of these scapes is plugged into and inextricably linked with the histories of the others—in the same way that, as we saw in the archaeological examples above, our media-infrastructural histories are deeply networked with our urban and architectural histories.

Richard John suggests that the "concept of an information infrastructure [for instance] . . . highlights the fact that the transmission of information has long been coordinated by a constellation of institutions, rather than by a single government agency or business firm."[38] As mentioned above, that infrastructural constellation includes not only institutions but also the everyday practices of ordinary people. It is important to recognize the codependency, the intertwining of these various entities and systems—the telegraph and the telephone, the railroad and the telegraph, transportation infrastructures and the postal system, print and writing infrastructures, writing and oral address, architecture and inscription, and various social and regulatory systems—and perhaps write their histories *together*.

Edwards, Jackson, Bowker, and Knobel lay out a general framework for how these "constellations" might form—in the cyberinfrastructure world, at least. It begins with system building; then technology transfer across domains; the emergence of variations in the original system design and the appearance of competing systems; the eventual merger of these various systems via *gateways*, into *networks*; the standardization of these networks and their merger into *inter-networks*—with, all the while, "early choices constrain[ing] the options available moving forward."[39] Such a model might seem rather deterministic to those of us looking at technology from a humanities orientation, or to those of us who are constructivists—yet I think this model identifies several phases, or pivot points, that occur during the maturation of technological systems that we already recognize and should be encouraged to look for. As Edwards, et al. suggest, "modeling" the formation of these networked infrastructural "constellations" does not imply that they are rigidly interlocked systems:

> The eventual growth of complex infrastructure and the forms it takes are the result of converging histories, path dependencies, serendipity, innovation, and "bricolage" (tinkering). Speaking of cyberinfrastructure as a machine to be built or a technical system to be designed tends to downplay the importance of social, institutional, organizational, legal, cultural, and other non-technical problems developers always face.[40]

These myriad infrastructures need to be networked into our media-infrastructural historiography. It is also important to situate those networked histories within the *longue durée*—to recognize, as John does, that systems and institutions have "long been coordinated" into an information infrastructure; or, as Edwards, et al. indicate, that their constellations are the result of "converging histories." So, rather than simply examining the intertwined technical, social, institutional, and cultural systems that gave rise to, say, cyberinfrastructure, we could acknowledge that *this* particular information infrastructure is networked into the long history of information infrastructures. Information itself has a deep time, as intellectual historians and library scientists have revealed.[41]

PATH DEPENDENCY

Path dependency, which Edwards et al. reference, is a particularly useful concept for scholars who have been taught to avoid at all costs being labeled a "techno-determinist," which, as Geoffrey Winthrop-Young jokes, "is a bit like saying that [one] enjoys strangling cute puppies."[42] Such suspect thinking often surfaces in "smart cities" rhetoric. There, the city, typically built *tabula rasa*, is equated with its technological infrastructure; the digital network *is* the city. Yet few live in cities that are born overnight; most metropoles are the product of decades, centuries, or millennia of expansion and renovation, razing and rebuilding, infilling and layering. In thinking about how these layers interact, humanities scholars often, in our overcompensation to avoid the scarlet *TD*, resist acknowledging the existence of well-trodden paths and how they have limited future choices. We see such paths in the long-term evolution of cities' media infrastructures. Architectural historian Kazys Varnelis offers a concrete example of paths' potency: "New infrastructures do not so much supersede old ones as ride on top of them, forming physical and organizational palimpsests—telephone lines follow railway lines, and over time these pathways have not been diffused, but rather etched more deeply into the urban landscape."[43] Thus it is possible to trace those infrastructural "paths" back into deep history. Doing so compels the recognition that those spaces built to accommodate historic forms of communication also *inform* and function as part of today's media infrastructures. The conceptual model of path dependency balances a recognition that technologies have material effects—that the channels laid and spaces configured by preceding technologies *do* steer the development, to some degree, of successor technologies—with an acknowledgment of the roles played by serendipity and tinkering, by historical social and cultural factors, in technological development.

PEOPLE AS INFRASTRUCTURE

The historical material record shows that people have not been mere beneficiaries of infrastructures but have actually served as infrastructures themselves. If, for instance, the public water supply does not extend into a particular neighborhood, residents of that neighborhood will often fill up their tanks and buckets within the service zone and tote their water that "last mile" home. People, in other words, do the work of absent pumps and pipes. This has been the case for millennia. There are plenty of parallels in media infrastructure. For instance, as Greg Downey has compellingly argued, messenger boys were a central link in the telegraph network.[44] In ancient Rome, as Johanson explained, residents transformed every surface of the built environment as a substrate for writing, and people used their voices to turn the volumes and surfaces of ancient cities into resonance chambers for public address. If important public notices were not distributed to peripheral urban zones, residents of those areas would bring *themselves* into the city center to hear or read the news. And as AbdouMaliq Simone argues, even today in Africa—and, undoubtedly, in much of the Global South and throughout much of global history—people often compensate for "underdeveloped, overused, fragmented, and often makeshift urban infrastructures."[45] The "incessantly flexible, mobile, and provisional intersections of residents . . . operate without clearly delineated notions of how the city is to be inhabited and used"—and they themselves fill in where their wires and pipes fall short.[46] Looking through the *longue durée* at the role *people* have played in infrastructural constellations helps us to appreciate the deeply entrenched and continuing centrality of biopower and human intellectual labor in our infrastructural constellations—"automated," digital, or otherwise.

INFORMAL / SHADOW DEVELOPMENT

Simone's mention of the flexible, mobile, and provisional suggests that infrastructure history—and media history in general—has been deeply informed by informal and "shadow" developments. In many parts of the developing (and even developed) world, where institutions do not provide, and perhaps have never provided, universal access to public services like media, islands of access within seas of exclusion are the norm. This is when people typically "go rogue." Brian Larkin writes about the jury-rigging, repurposing, or pirating of existing infrastructures in Nigeria. Such improvisations have appeared throughout media history—as in the cooptation of building facades as substrates for public writing in ancient Rome and Egypt—and these peripheral practices should factor into our media-infrastructural histories. Consider the long history of people making unofficial marks, graffiti, on urban walls; or the long history of pirated publication and urban shadow-markets for unauthorized texts; or the long history of people making unauthorized noise—proselytizing or hawking their wares—in public space.[47]

Thinking about the "deep time" of media infrastructure—back beyond those technological systems administered by modern states, governmental agencies, and multinational corporations—reveals that as infrastructures have become increasingly institutionalized, centralized, and networked, what constitutes "informality" has also evolved. Situating informal infrastructures in relation to the long history of infrastructure uncovers the fact that an infrastructure's "shadow" has a history too.

SCALE

In examining infrastructures of vocality and writing I have considered entities as small as the individual voice and as big as an entire urban form. Today's infrastructures, of course, encompass global networks and even extraterrestrial domains. Infrastructures thus compel thinking about the granularity of observations; Graham and Marvin list various scales of infrastructural analysis, including the corporeal, local, urban, regional, national, international, and global.[48] When writing media-infrastructural histories, it matters whether one is writing media object histories, local media histories, urban media histories, national media histories, or cultural media histories, and making a choice between them can be complicated by the fact that infrastructures extend across these scales, connecting technologies into networks and internetworks. Paul Edwards suggests that scale need not be conceived of as merely a geographic quality; it is also possible to consider scales of force (from the human body to the geophysical), scales of time (from human time to geophysical time), and scales of social organization (from individuals to governments).[49] Again, infrastructures span all these scales. And those scales—what constitutes the "nation" or how one conceives of the boundaries of the "subjectivized" body, for instance—also have a deep history.

The macro spatiotemporal view is particularly illuminating in that it forces consideration of the forms of media and infrastructures in relation to their long-term functions—"the reasons they came to exist in the first place."[50] Rather than thinking about how the telegraph supplanted the postal service, or how writing supplanted the voice, for instance, these two systems can be thought of as two instantiations of a shared infrastructural *purpose*. As Edwards suggests, contextualizing the telephone, the telegraph, the post, and other modern technologies within James Beniger's "control revolution" concept "allows us to understand not only the genesis and growth of the many large infrastructures that characterize modernity, but also the process of linking these infrastructures to each other."[51] Of course, we would need to identify alternative infrastructural *purposes* to encompass our premodern infrastructures, too. Whatever those purposive thematics or ideologies might be, this act of linking and contextualizing foregrounds the historical continuity (and perhaps some discontinuities) among infrastructures—the long now, the "deep time"—and the myriad structures that have intertwined in order to allow us to traffic in signals of myriad forms across the ages.

Notes

1. Clifford Geertz, *Negara: The Theatre State in Nineteenth-Century Bali* (Princeton, N.J.: Princeton University Press, 1980); Peter Hall, *Cities in Civilization* (New York: Pantheon, 1998); Paul Wheatley, *The Pivot of the Four Quarters: A Preliminary Enquiry into the Origins and Character of the Ancient Chinese City* (Chicago: Aldine, 1971).

2. Lewis Mumford, *The Culture of Cities* (1938; repr., New York: Harcourt Brace Jovanovich, 1966), 6.

3. Siegfried Zielinski, *Deep Time of the Media: Toward an Archaeology of Hearing and Seeing by Technical Means* (Cambridge, Mass.: MIT Press, 2006). In their 2007 NSF-funded workshop on cyberinfrastructure, Paul Edwards and several colleagues argued for the importance of studying the "long now" of cyberinfrastructure: the two hundred years' worth of "slower-pace[d]" political, cultural, and technical changes that have been happening "in the background"—changes like the rise of scientific disciplines and statistics—that have laid the foundation for digital networks (Paul N. Edwards, Steven J. Jackson, Geoffrey C. Bowker, and Cory P. Knobel, "Understanding Infrastructure: Dynamics, Tension, and Design," workshop, "History and Theory of Infrastructure: Lessons for New Scientific Cyberinfrastructures," University of Michigan, Ann Arbor (January 2007), 3. Of course I would argue that media studies could benefit from a much longer view, one that recognizes that "infrastructure" precedes the "cyber" and the electronic—but still, these scholars' focus on historical contextualization is useful. And the concept of the "long now," a contemporary that extends into the past, complements Zielinski's call to "find something new in the old"—to find the "now" in

history—and recent efforts to consider the "geology" of media: the natural resources used to make our media hardware. See Jussi Parikka, "Deep Times and Geology of Media," *Machinology*, August 20, 2013, available at http://jussiparikka.net/2013/08/20/ deep-times-and-geology-of-media (accessed September 20, 2014). These latter efforts are particularly relevant to our efforts to dig into deep history, given that the very concept of deep time emerged in geology.

4. Paul Edwards, "Infrastructure and Modernity: Force, Time, and Social Organization in the History of Sociotechnical Systems," in *Modernity and Technology*, ed. Thomas J. Misa, Philip Brey, and Andrew Feenberg (Cambridge, Mass.: MIT Press, 2003), 186–87, 188.

5. Raymond Williams, *Marxism and Literature* (New York: Oxford University Press, 1977), 122.

6. James Ackerman, *Origins, Imitations, Conventions: Representation in the Visual Arts* (Cambridge, Mass.: MIT Press, 2002); Beatriz Colomina, *Privacy and Publicity: Architecture as Mass Media* (Cambridge, Mass.: MIT Press, 1994); Peter Bacon Hales, *Silver Cities: Photographing American Urbanization, 1839–1915* (Philadelphia: Temple University Press, 1983); Levine, Neil. " 'The Significance of Facts': Mies's Collages Up Close and Personal," *Assemblage* 37 (December 1998): 70–101; Richard Pare, *Photography and Architecture: 1839–1939* (Montreal: Canadian Center for Architecture, 1982); Shelley Rice, *Parisian Views* (Cambridge, Mass.: MIT Press, 1999).

7. Walter Benjamin, *Illuminations*, trans. Harry Zohn (New York: Schocken, 1969); David B. Clarke, ed. *The Cinematic City* (New York: Routledge, 1997); Colomina, *Privacy and Publicity*; Edward Dimendberg, *Film Noir and the Spaces of Modernity* (Cambridge, Mass.: Harvard University Press, 2004); Sergei Eisenstein, "Montage and Architecture," trans. Michael Glenny, *Assemblage* 10 (1938/1989): 111–31; Scott McQuire, *The Media City: Media, Architecture and Urban Space* (Thousand Oaks, Calif.: Sage, 2008).

8. David Heckman, *Small World: Smart Houses and the Dream of the Perfect Day* (Durham, N.C.: Duke University Press, 2008); Anna McCarthy, *Ambient Television: Visual Culture and Public Space* (Durham, N.C.: Duke University Press, 2001); Scott McQuire, Meredith Martin, and Sabine Niederer, eds, *Urban Screens Reader*, Institute of Network Cultures Reader, vol. 5 (Amsterdam: Institute of Network Cultures and Creative Commons, 2009); Lynn Spigel, *Make Room for TV: Television and the Family Ideal in Postwar America* (Chicago: University of Chicago Press, 1992).

9. Karin Bijsterveld, *Mechanical Sound: Technology, Culture, and Public Problems of Noise in the Twentieth Century* (Cambridge, Mass.: MIT Press 2008); Steve Goodman, *Sonic Warfare: Sound, Affect, and the Ecology of Fear* (Cambridge, Mass.: MIT Press, 2010); Brian Larkin, *Signal and Noise: Media, Infrastructure, and Urban Culture in Nigeria* (Durham, N.C.: Duke University Press, 2008); Emily Thompson, *The Soundscape of Modernity: Architectural Acoustics and the Culture of Listening in America, 1900–1933* (Cambridge, Mass.: MIT Press, 2004); Shundana Yusaf, *Broadcasting Buildings: Architecture on the Wireless, 1927–1945* (Cambridge, Mass.: MIT Press, 2014).

10. McQuire, *Media City*, vii.

11. Eric Gordon, *The Urban Spectator: American Concept-Cities from Kodak to Google* (Hanover, N.H.: Dartmouth College Press, 2010), 2.

12. Mario Carpo, *Architecture in the Age of Printing: Orality, Writing, Typography, and Printed Images in the History of Architectural Theory* (Cambridge, Mass.: MIT Press, 2001). See also Diane Favro, "Meaning and Experience: Urban History from Antiquity to the Early Modern Period," *Journal of the Society of Architectural Historians* 58, no. 3 (1999): 364–73; Rose Marie San Juan, *Rome: A City Out of Print* (Minneapolis: University of Minnesota Press, 2001); Bronwen Wilson, *The World in Venice: Print, the City, and Early Modern Identity* (Toronto: University of Toronto Press, 2005).

13. See also Shannon Mattern, "Paju Bookcity: The Next Chapter" *Places*, January 14, 2013, available at https://placesjournal.org/article/paju-bookcity-the-next-chapter (accessed September 20, 2014); and "Interfacing Urban Intelligence" *Places*, April 28, 2014, available at https://placesjournal.org/article/interfacing-urban-intelligence (accessed September 20, 2014).

14. Shannon Mattern, "Ear to the Wire: Listening to Historic Urban Infrastructures" *Amodern* 2 (Fall 2013), available at http://amodern.net/article/ear-to-the-wire (accessed September 20, 2014).

15. Quintilian, *Institutio Oratoria* (ca. 95 CE), 2.16.9, available at http://perseus .uchicago.edu/perseus-cgi/citequery3.pl?dbname=LatinAugust2012&getid=1& query=Quint.%202.16.15 (accessed September 20, 2014); See also Indra Kagis McEwen, "Hadrian's Rhetoric I: The Parthenon," *RES: Anthropology and Aesthetics* 24 (1993): 55–66.

16. Aristotle, "Politics," in *Complete Works of Aristotle*, Revised Oxford Translation, ed. Jonathan Barnes (Princeton, N.J.: Princeton University Press, 1998), 1326b5–7; Vitruvius, *The Ten Books on Architecture* (Cambridge, Mass.: Harvard University Press, 1914), 139.

17. Diane Favro and Christopher Johanson, "Death in Motion: Funeral Processions in the Roman Forum," *Journal of the Society of Architectural Historians* 69, no. 1 (2010): 15.

18. Christopher Lyle Johnstone, "Communicating in Classical Contexts: The Centrality of Delivery," *Quarterly Journal of Speech* 87, no. 2 (2001): 121–43.

19. David Henkin, *City Reading: Written Words and Public Spaces in Antebellum New York* (New York: Columbia University Press, 1998).

20. Joanna Merwood-Salisbury, "Patriotism and Protest: Union Square as Public Space, 1832–1932," *Journal of the Society of Architectural Historians* 68, no. 4 (2009): 543.

21. Ibid., 551.

22. Christopher Johanson, interview with the author, February 26, 2013.

23. Louise Revell, *Roman Imperialism and Local Identities* (New York: Cambridge University Press, 2009), 3–4.

24. Irene A. Bierman, *Writing Signs: The Fatimid Public Text* (Los Angeles: University of California Press, 1998), 20.

25. Robert E. Harrist Jr., *The Landscape of Words: Stone Inscriptions from Early and Medieval China* (Seattle: University of Washington Press, 2008), 23.

26. Harold Innis, *The Bias of Communication* (Toronto: University of Toronto Press, 1951).

27. Brinkley Messick, *The Calligraphic State: Textual Domination and History in a Muslim Society* (Berkeley: University of California Press, 1993); Mumford, *Culture of Cities*.

28. Messick, *Calligraphic State*, 231.

29. Ibid., 246–7.

30. See London School of Economics, *Electric City* (London: LSE, December 2012), available at http://lsecities.net/publications/conference-newspapers/the-electric-city, which includes work by several key "smart city" critics, including Orit Halpern, Dan Hill, Saskia Sassen, and Richard Sennett. See also *Volume* 34 "City in a Box" Special Issue (December 2012); Adam Greenfield's *The City is Here for You to Use* series of e-books; and Shannon Mattern, "Methodolatry and the Art of Measure" *Places* (November 5, 2014), available at https://placesjournal.org/article/methodolatry-and-the-art-of-measure (accessed September 20, 2014).

31. Richard Sennett, "No One Likes a City That's Too Smart" *Guardian*, December 4, 2012.

32. Henri Lefebvre, *The Production of Space*, trans. Donald Nicholson-Smith (1974; Oxford: Blackwell, 1991).

33. Larkin, *Signal and Noise*, 5.

34. Mattern, *Ear to the Wire*, 2013.

35. Stephen Graham and Simon Marvin, *Telecommunications and the City: Electronic Spaces, Urban Places* (New York: Routledge, 1996), 329.

36. Christopher L. Witmore, "Vision, Media, Noise, and the Percolation of Time: Symmetrical Approaches to the Mediation of the Material World," *Journal of Material Culture* 11, no. 3 (2006): 279.

37. Stephen Graham and Simon Marvin, *Splintering Urbanism: Networked Infrastructures, Technological Mobilities and the Urban Condition* (New York: Routledge, 2001), 8.

38. Richard R. John, "Recasting the Information Infrastructure for the Industrial Age," in *A Nation Transformed: How Information Has Shaped the United States from Colonial Times to the Present*, ed. Alfred D. Chandler and James W. Cortada (New York: Oxford University Press, 2000), 56.

39. Edwards et al., "Understanding Infrastructure," i–ii.

40. Ibid., 6–7.

41. See Ronald E. Day, *The Modern Invention of Information: Discourse, History, and Power* (Carbondale: Southern Illinois University Press, 2001); James Gleick, *The Information: A History, a Theory, a Flood* (New York: Vintage, 2012); University of California, San Diego Global Information Industry Center's "History of Information," available at http://giic.ucsd.edu/historyofinfo.php; or the syllabus for Shannon Mattern, "Archives, Libraries and Databases," graduate seminar, available at http://www.wordsinspace.net/lib-arch-data/2014-fall.

42. Geoffrey Winthrop-Young, *Kittler and the Media* (Malden, Mass.: Polity, 2011), 121.

43. Kazys Varnelis, "Centripetal City," *Cabinet* 17 (Spring 2005): 27–28.

44. Gregory Downey, *Telegraph Messenger Boys: Labor, Technology, and Geography, 1850–1950* (New York: Routledge, 2002). See also Gregory Downey, "Making Media Work: Time, Space, Identity, and Labor in the Analysis of Information and Communication Infrastructure," in *Media Technologies: Essays on Communication, Materiality and Society*, ed. Tarleton Gillespie, Pablo J. Boczkowski, and Kirsten A. Foot (Cambridge, Mass.: MIT Press, 2014), 141–65.

45. AbdouMaliq Simone, "People as Infrastructure: Intersecting Fragments in Johannesburg," *Public Culture* 16 (September 2004): 425.

46. Ibid., 407.

47. See Adrian Johns, *Piracy: The Intellectual Property Wars from Gutenberg to Gates* (Chicago: University of Chicago Press, 2010); Jennifer A. Baird and Claire Taylor, *Ancient Graffiti in Context* (New York: Routledge, 2011); Hillel Schwartz, *Making Noise: From Babel to the Big Bang and Beyond* (Brooklyn: Zone, 2011).

48. Graham and Marvin, *Splintering Urbanism*, 411.

49. Edwards, "Infrastructure and Modernity," 186.

50. Ibid., 204.

51. Ibid., 207.

PART II

Resources, Environments, Geopolitics

CHAPTER 5

Water, Energy, Access

Materializing the Internet in Rural Zambia

LISA PARKS

nfrastructure is both the thing and the story. It is the transparent and the spectacular. It is seamless in its operation and can be disastrous in its failure. It is something we do not know whether we should want and something we think we cannot live without. It is what tethers us together and what sets us apart. This chapter explores the material conditions of Internet infrastructure in the rural community of Macha, Zambia. Located in the country's southern province, Macha is home to 135,000 Tonga people, many of whom speak Chitonga and English, the country's national language since 1964. The community is the site of a Brethren Church of Christ mission, a regional hospital and malaria research institute, several mission-run and state-supported schools, a small, open-air market, and an information technology academy, all of which are connected to the nation's electrical grid. Most Machans, however, live off the grid in small, scattered homesteads in extended families (see figure 5.1). Many are subsistence farmers who live on an average income of less than one dollar per day.[1] The nearest grocery store is seventy kilometers away in a city called Choma, and it costs five dollars to ride there on a minibus. It costs thirty dollars per month for a voucher to access the Internet.

In 2012 and 2013, I conducted fieldwork in Macha as part of a collaborative research project that involved partnering with community members to design sustainable Internet and mobile phone systems scaled to rural socio-economic conditions and informed by local needs, interests, and desires. This aspiration

Figure 5.1. Most Machans live in homesteads scattered around the center of this rural Zambian community. Photo by author.

to work closely with community members as part of the process of integrating information and communication technologies (ICTs) in rural areas arises on the heels of other ICT for development (ICTD) projects that have failed because of limited engagement with people who would be using new technologies.[2] Much ICTD work is underpinned by development ideology—a blind faith in the capacities of ICTs to "modernize" and "enhance" the lives of anyone fortunate enough to come within their reach. This development ideology sets the tone for many ICTD projects seeking to address the needs of the so-called "O3B" or "other 3 billion"—the mass of people still without Internet access who are alternately imagined as a technologically disenfranchised class or a giant untapped market.[3] In the context of such logics, many Africans feel their communities have either become test sites for Westerners doing feel-good ICT research or a dumping ground for the West's digital hand-me-downs—old computers and printers shipped to Africa, many of which are obsolete, broken, or incompatible with local electrical systems, and thus useless.[4] In Macha an entire cargo container sent from Europe and filled with computer equipment sat unused for months. After the container became infested with termites,

Figure 5.2. Machans could access the Internet from 2011 to 2012 from several LinkNet VSAT terminals built out of repurposed cargo containers. Photo by author.

much of its contents were burned, causing exposure to toxic incineration of plastic and metal parts. Other donated computers have piled up in storage rooms awaiting software installations.

Internet access was first established in Macha in 2004 through a VSAT system at the community's Malaria Institute at Macha (MIAM), run by Johns Hopkins University. Shortly thereafter, a local organization called LinkNet formed, and, with the support of the Dutch government, developed rural Internet services by installing cargo containers equipped with solar panels, VSAT systems, and computers in village sites, charging access fees with a voucher system (see figure 5.2). Access charges ranged from three kwachas (sixty cents) for a few minutes access to 150 kwachas (thirty dollars) for a month. These voucher revenues barely made a dent in the high-cost satellite gateway to the Internet, which ran up to seven thousand dollars per month and was subsidized by Dutch organizations and the UK company AfriConnect.[5] To expand the user base and generate more cost-sharing revenue, in 2009 LinkNet began installing a Wi-Fi mesh network in Macha that would enable it to collect service fees from schools, churches, the hospital, the radio station, and private residences.

After this change to the infrastructure, which was designed to extend access and generate revenue, financial problems for LinkNet persisted.

By 2012 LinkNet and its umbrella Macha Works[6] went bankrupt, and a clear digital divide had formed in Macha related to several factors. First, the community's early adopters were situated within the village's center, which was not only connected to the electrical grid but also to historically colonial institutions such as the hospital, the mission, and the schools. There has been little sustained ICT education of or outreach to people in the thirty-five-kilometer radius beyond Macha's center. Those who had familiarity with computing had either taken an International Computer Driver's License (ICDL) course offered by LinkNet's IT academy (LITA) or received ICT training outside of Macha. Second, most Machans do not know about computing or the Internet. Since Tonga is an oral culture, many have not learned to read or write in English, much less how to type or use a mouse, and therefore they have little interest in or incentive to adopt computer technology. Compounding this issue is the fact that the local indigenous language—Chitonga—was not used in LinkNet's Internet access instructions or on the Macha Works website. Third, most Machans, even those who knew about computers and the Internet, could not afford Internet vouchers or computer equipment and did not have electricity in their homes. Machans who wanted to access the Internet tended to do so opportunistically, either through their workplace or by looking over someone's shoulder. Fourth, as LinkNet was trying to establish Internet service in Macha and attract users, two commercial mobile phone providers, Airtel and MNT, managed to do so with more affordable pricing options through mobile phones, a much more familiar and readily accessible technology among Machans. Finally, in 2012 serious conflicts surfaced among the leaders of Link Net and Macha Works and the community's traditional stakeholders, an amalgam of local chiefs, leaders of the Brethren Church in Christ mission, Tonga spiritualists, and community members. These conflicts resulted in allegations of embezzlement and corruption, leading to litigation, extradition, and ongoing strife in the community. In the fall of 2012 Tonga sorcerers reportedly cast spells on former LinkNet employees, causing some to flee Macha and seek refuge in neighboring Zimbabwe.

To be clear, I am not an anthropologist or an Africanist. I am a media and communication scholar interested in the arrangement and use of ICTs, and since our fieldwork began in 2012, navigating through and making sense of these conflicts and shifting conditions up close and from afar has been challenging. I was asked to join this project because I had conducted fieldwork and ethnographically inspired research on uses of media technologies in rural communities in other parts of the world—specifically on Aboriginal Australian uses of satellites and Mongolian uses of mobile telephony.[7] In addition, I have tried to develop critical materialist approaches for studying media infrastructures—the biophysical resources, designs, hardware, and labor organized to support the circulation of signal traffic across different cultural and territorial contexts.[8] Our ICT research in Macha continues this work by investigating the players, resources, and power struggles involved in the installation and use of Internet infrastructure in this rural African community. The focus is on formulating relational and hybrid understandings of the arrangement and use of ICTs across cultural contexts, not on the mastery of Others' cultures. In this way, the project builds upon the work of Eric Michaels, Faye Ginsberg, Brian Larkin, Jenna Burrell, and others who have explored how people in developing and postcolonial contexts have adapted or "reinvented" imported technologies, localizing and using them to contest Western hegemony, create tactics of cultural survival, or respond to oppressive state policies or socioeconomic conditions.[9]

So what I propose to explore here is the materialization of Internet infrastructure in rural Zambia, the technical facilities that enable local links to the global Internet as well as the natural resources, sociotechnical relations, and institutions that are organized to sustain these facilities and links. Since many Machans use mobile phones to access the Internet, I consider mobile telephony as part of their Internet infrastructure. By "materialization" I am referring to how "matter becomes." Studying materialization involves recognizing the constitution of phenomena as part of a "multitude of interlocking systems and forces," grasping the complex dynamics of causation, and tracking the "changing location and nature of capacities for agency."[10] Such study reveals the micro- and macro-level forces, contingencies, and conflicts that can inform and result from Internet infrastructure's emergence in specific locations and their relation to everyday culture.

The chapter begins with a discussion of the resource requirements of ICTs in Macha, drawing out the contingent relations between information, water, and electrical systems. Then, to convey ICT imaginings and uses within these conditions, I present a sampling of Machan Internet-access stories, focusing on the relation between Internet infrastructure and local agriculture, transportation, and gender politics. My fieldwork in 2012 and 2013 included site visits in and around Macha, videotaped interviews with nearly two hundred community members, group meetings, informal conversations, photography, and the installation and testing of an experimental wireless network called VillageCell. This research revealed that Internet infrastructure is inseparable from the electricity and human biopower that energizes it, the layering of systems that precede it (such as water, agriculture, transportation), and the multifarious ways people imagine, use, and respond to it. Understanding the materialization of Internet infrastructure in rural Zambia works to destabilize dominant discourses that posit ICT diffusion and adoption in rural Africa as a straightforward path to "modernization," "development," and "global integration," and instead points to local political, economic, and cultural challenges to the Internet's globalization.

Energizing ICTs in Rural Zambia

It is impossible to think about or use the Internet or mobile phones in Macha (or perhaps anywhere) without thinking about water. In Zambia, hydropower is the primary source of electricity. Hydropower plants were built during the period of British colonial rule in the early twentieth century, in what was then Rhodesia, to support the copper mining industry in the northern part of the country, now known as the Copper Belt.[11] Today, hydropower plants in Kafue Gorge, Kariba North Bank, and Victoria Falls generate most of the country's electricity, which is networked on a grid administered by ZESCO—the Zambian Electrical Supply Company—and is connected to South Africa, Democratic Republic of Congo, and Zimbabwe. The grid also interlinks major Zambian cities and a handful of rural villages (especially those with hospitals or important facilities) and is used to export electricity to Botswana, Tanzania, and Namibia. Even though Zambia has a surplus of electricity from its hydropower resources, power goes on and off unpredictably almost every day in big cities and small villages alike. There are fluctuations and inconsistences in the voltage, which can cause damage to computers and other electronics—not to mention fires—and power outages regularly interrupt Internet and mobile phone services. Other common causes of Internet downtime in Zambia are adverse weather conditions, low or shared bandwidth, and poor quality of copper cables and telephone connections.[12]

Rural communities such as Macha are particularly vulnerable to load shedding, the centralized practice of shutting off services to one area to support demand in another, often to the advantage of urban areas or clients abroad. Almost every day the power goes out in the community at unpredictable times, sometimes for a few minutes but often for a few hours. The graph in figure 5.3 shows the irregularities and fluctuations in power use over two weeks in Macha.[13] Most Machan Internet users I interviewed expressed frustration with this situation and described their use as punctuated by frequent disruption. Such conditions foreground the uneven temporalities of networks and experiences of Internet connectivity. Far from being a universal clock, homogeneous in its durations, the Internet's dynamism is contingent on the harnessing of energy

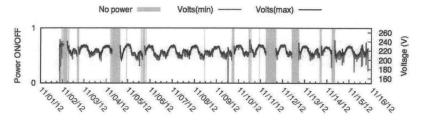

Figure 5.3. This graphic shows power voltage variations over a two-week period in Macha during 2012. The grey vertical bands represent power outages.

resources, the timed flow of electrical currents, and the regulation of voltage. While, as Tiziana Terranova reminds us, the Internet cannot be reduced to a grid or a database, its dynamism relies on steady access to electrical power.[14]

The integration of Internet and mobile-phone infrastructures within daily lives depends on the conversion of water, sun, fossil fuels, and other materials into electricity. In *Greening the Media*, Rick Maxwell and Toby Miller discuss the growing electrical demands of the global digital economy, noting, "By 2011, upwards of ten billion devices needed external power supplies, including two billion TV sets, a billion personal computers, and [six billion] cell phones."[15] Not only is the Internet powered by multiple energy sources, it is increasingly used in combination with sensors, cameras, and instruments to remotely monitor and manage the world's energy resources and extraction sites, whether oil pipelines, hydropower facilities, or nuclear power plants. Because of this, there is an urgent need to explore energy resource requirements as part of critical discussions of global signal traffic—that is, there is a need *to consider the external, material demands of information infrastructures in tandem with their internal dynamics, logics, forms, and cultures.*

In Zambia it is the movement of water through the country's sovereign territory, down rivers and into the mouths of hydropower plants that generates the capacity to access the Internet and use mobile phones. Having said this, it is important to point out that most Machans live "off the grid." It costs seventy-five thousand kwachas (fifteen thousand dollars) to electrify a home.[16] Given this high cost, most Machans power their devices in other ways. Some use power outlets at the outdoor market in the village center to charge their electronics and pay a fee to do so. Some use solar panels to power up radios, lights, and other small appliances. And some jerry-rig car batteries to energize TV sets and stereo systems. Consumer electronics are fueled not only by hydropower but also by the manual labor of people who take time and energy each day to devise

Figure 5.4. This sign appeared at Macha's open-air market, where people from the area come to charge mobile phones and other electronics. Photo by author.

ways of powering their devices and by local autodidacts who have figured out how to repair and maintain them.

Just off the dirt road leading to Macha's center, a local repairman works atop a small patch of cement outside an abandoned shop where people from the area drop off broken generators, radios, power converters, and other objects, hoping they can be repaired. The objects are arrayed around him as he kneels, crouches, or sits on the ground and works with a few coveted tools. The repairman cracks things open, pulls the pieces apart, lays them on pieces of cloth, tries to identify the problem, figures out a solution, finds or makes needed parts, and puts machines back together in an effort to extend their lives. Radios or mobile phones are fixed alongside and sometimes in relation to other objects such as small engines, bicycle wheels, or water pumps, which support daily life in this agricultural community. Skill and knowledge aggregate through regular and rigorous tactile engagements with different kinds of technical objects rather than via mastery over one thing. When there is a conundrum in repairing one object, another is tackled, and the solution for that repair may inform others.

As Steve Jackson suggests, such acts of repair are articulated with technological diversity, the circulation of knowledge/power, and the ethics of care.[17] Repair work also creates solutions for keeping machines and systems running in areas where energy, funding, and commodity flows are limited.

In addition to regular power failures in Macha, the community's water supply is limited and uneven. Though there is a central water tower in Macha (see figure 5.6), along with a few wells, most homes do not have modern plumbing or running water. Instead, there are spigots placed in sites throughout the community and, as in many other parts of the world, people (often women or children) convene around them and fill up large buckets or containers (when the water is flowing) and carry them home. Like electricity, sometimes water flows in Macha, and sometimes it ceases altogether, often at unpredictable times. As one informant explained, "My wife has to fetch water. You'll find that she'll go there for an hour or two, just to bring water. It's very often that maybe there'll be a queue, because of water delays, and that she could pump for eight minutes before water would come out. Or we will find that the whole day there won't be, there won't be water. So it's truly a challenge in Macha." While we were in Macha in 2012, our spigot was dry for three days. On such occasions, community volunteers fill large containers at the water tower's spillover basin or at a functioning spigot and haul them via tractor to people's scattered homes. Without water there could be no Internet or mobile phone users in this area. People need water to live; they need to be alive to use Internet and mobile phone infrastructure.

Put another way, *this* water infrastructure (see figure 5.5a)—the movement of water performed by Machan women—supports *this* Internet access (see figure 5.5b)—children in a local private school. Machan water carriers not only support and sustain their families but, in the logic of digital capitalism, their labor is implicitly commandeered to sustain populations on the cusp of becoming new markets for commercial Internet service providers and mobile telephone companies seeking to extend their enterprises into new regions as market saturation peaks in industrialized parts of the world. Reinforcing this point, an ad for Airtel, one of Zambia's most popular mobile phone networks, features an image of a prospective subscriber in a rural village setting and proclaims, "If you're out there, we will find you."

Relations between water and Internet infrastructure in Macha are materialized in another way as well. An array of transmitters has been mounted on the community's central water tower, which, at thirty meters, is one of the tallest points in the area (see figure 5.6). The water tower hosts Macha's community

Figure 5.5b. School children use computers and access the Internet in a private Christian school in Macha. Photo by author.

Figure 5.5a. Women's bodies and labor are integral parts of water distribution infrastructure in Macha.

radio transmitter as well as masts and antennas installed by LinkNet to extend the community's Internet infrastructure. Here water and signals transit through the same node. Internet infrastructure is quite literally supported by water infrastructure. The digital economy is layered upon the resource economy. Our research team contributed to this tower archaeology when we installed antennas and a base station to test an experimental mobile phone network called VillageCell, which uses the white space part of the electromagnetic spectrum to provide free local mobile-phone service in Macha.[18] The creation of telecommunication infrastructure, which is often done relatively invisibly by linemen and tower workers in urban and rural areas alike, is a biophysical process that involves assembling and hauling equipment, climbing up and down towers, measuring positions, making adjustments, and performing tests. After numerous challenges, our team finally managed to get VillageCell operating—only to have its base station struck by lightning and out of commission a few months later. Infrastructure development is a process that takes time and involves failure.

One motivating factor for our infrastructure research in Macha was the recognition that if an Internet user in the community wanted to send an email or share a photo with another person in Macha, the data would have to be routed

Figure 5.6. Macha's water tower hosts several antennas and was one of the sites of our team's VillageCell experiment. Photo by author.

up via satellite to servers thousands of miles away in Silicon Valley and then back down the satellite link through servers in Macha. As already mentioned, the community and its Dutch supporters were paying up to seven thousand dollars per month for the capacity to route data intended for local exchange through servers on other continents.[19] If this were not enough, the satellite Internet gateway experienced latency, slow speeds due to limited bandwidth, and frequent downtimes because of power outages. The technical portion of our research thus involved designing systems to support local control of data exchange and reduce cost, congestion, and Internet downtime.[20] Because of high gateway fees, it is five times more expensive for Zambians to access the Internet in rural areas than in urban areas.[21] In an effort to help alleviate high gateway costs, in 2011 AfriConnect provided a microwave Internet link in Macha, reducing LinkNet's monthly expense to four thousand dollars, which still proved too costly. The completion of a ten-thousand-kilometer transoceanic cable between Sudan and South Africa, called EASSy (which cost $263 million), is expected to provide Zambia with a fiber-optic Internet link that will further reduce Internet access costs in the country's interior.[22]

Imagining and Accessing the Internet

In addition to considering the natural and human resources that support information infrastructure, our research in Macha explores how people imagine and use the Internet and mobile-phone technologies in the community. To develop this part of our work, we devised a collaborative ethnographic method that involved close partnership with twenty Machans.[23] First, we discussed a preliminary list of research topics and questions and then invited our partners to alter or add to it. Second, we trained our partners to use Flipcams and record interviews. Third, we asked them to conduct videotaped interviews of Machans in and beyond the community's center about their Internet and mobile phone use. Together our team conducted 178 videotaped interviews (some of which included multiple people) ranging from five to ninety minutes in length. A more extensive analysis of our ethnographic methodology and findings is developed in another publication.[24] For now, I want to provide a general overview of our findings and present some exemplary comments about ICT use in Macha.

Of the 135,000 people in Macha, approximately three hundred regularly used the Internet in 2012 via desktop or laptop computers. Those who access it do so through their work with the hospital or schools, and a select few have access at home. By 2013, after LinkNet had been shut down, a growing number of Machans were using their mobile phones to access the Internet. Most regular users reported familiarity with Google and Yahoo search engines and email services as well as Facebook and Skype, and indicated they use the Internet to seek information related to their work in fields of education, healthcare, and community development, and to read national and international news. Those who use Facebook do so primarily to connect with locals; some use it to stay in touch with family and friends who live in other parts of Zambia or abroad.[25] Most Machans we interviewed felt the Internet enhanced their lives by improving access to information and connecting their community to a broader sphere of activity. A select few indicated a concern with access to pornography and wasting time. They expressed favorable feelings about the technology's potential, often reiterating ICT development ideology, but complained about interrupted, slow, and costly access. When asked if the Internet should be free, most replied no, acknowledging that there are overhead costs associated with

Internet services that must be covered, yet some pointed out that the Internet in Macha has become primarily a technology for the rich.

Those Machans who use ICTs to access the Internet had a range of opinions and experiences to share. One Machan man conveyed the sense of intimidation he felt when first encountering a computer, indicating, "Yeah, when I was there learning, the fear came out, even to hold the computer, it wasn't an easy thing for me." A teacher who uses the Internet to prepare for his classes noted the frustration of being disconnected because of frequent power outages: "You find maybe for three days people given the mandate to run the Internet will just make an apology: Internet is down, we are doing everything possible. So the running down of the Internet. Also, the opening of the Internet. Those are the major challenges." Another Machan expressed the feeling of isolation that arose after being connected and disconnected: "You know, when you are not online, or when you are not connected to the Internet, it's more like you're outside the world. So, it's important that each and everyone looking at the civilized world we are in today—each one should be connected to the Internet. Each one should be updated with Internet, to know how things are around us." A hospital administrator emphasized how reliant he has become on his mobile phone: "I think the mobile phone is the center of information . . . without it, it's like life without blood. To a human being. Because I'll be paralyzed. I'll be completely paralyzed without this technology. And that's how I view it, because sometimes it creates information gap, when people don't reach you, when you don't reach them." Finally, Chief Macha emphasized the unpredictable potentials of new technologies: "Any technology can be very useful, but it can be misused. That's why one must be very careful. These phones that we use are very, very useful actually. Very, very useful. But they can be very destructive. I could misuse this for to kill you. I can misuse this for to do anything stupid. So it is a question of how you use the device."

Beyond conveying a sense of fear, utility, dependence, and the unpredictable outcomes of ICT use, Machans discussed a variety of other topics during our interviews, ranging from education to mobility, from time to technological breakdowns, from community development to resource access, from entertainment to employment. Our interviews revealed many issues of concern, but here I focus on three that recurred in our discussions: use of the Internet to support local farmers; use of the Internet to purchase used cars from Japan; and a lack of awareness of or an indifference to the Internet. Addressing these issues will foreground how the Internet has been imagined and used in relation to other existing infrastructures of agriculture, transportation, and communication,

Figure 5.7. Farmers around Macha have created ways of distributing important details about crop prices via Internet, radio, and mobile phone as they move around their farms. Photo by author.

and will highlight the reality that many Machans live without ever accessing the Internet and may not want it at all.

Daisy-chaining Crop Prices

Most Macha families are subsistence farmers, and some run larger farms that grow maiz, sunflowers, soy, ground nuts, and cotton. Farmers have found it increasingly important to know the crop prices set by the Zambian National Farmers Union before taking their crops to market so that they are not cheated by so-called "briefcase buyers" who buy crops at the lowest possible price. Since not all farmers in the community have Internet access, a practice has emerged whereby one farmer with Internet access finds crop prices or agricultural news online or listens to the radio and then broadcasts the information via text message to farmers with mobile phones. In this case, Internet and mobile-phone systems are productively interlinked or "daisy-chained" by information elites in the community. Mobile phone networks are used to extend the flow of information via personal networks into areas that either do not have Internet access or do not have Internet users. The practice has also been performed by Macha's community radio station, which has operated since 2005 and reaches an area with a diameter of about 140 kilometers, covering at least four chiefdoms and potentially reaching 150,000 people.[26] Though the range of the radio station's coverage is expansive, crop prices are announced via radio only at fixed times during the day. Mobile phones allow the information to be circulated at the user's discretion. And farmers can receive information via text message as they move about their farms, whether they are outside or inside, or in transit to other areas (see figure 5.7). Similar cross-platform practices are beginning to occur in Macha with healthcare services as hospital workers are using mobile phones to push health alerts and immunization notices to people throughout the area.

Buying Used Cars from Japan

Several people we interviewed told stories about men in Macha using the Internet to buy used cars from Japan for two thousand to three thousand dollars via a website called Conjunction.com. The cars are shipped in cargo containers to Tanzania and then trucked overland across the border and driven across dirt roads into Macha. Interviewees expressed great pride and excitement about these major online purchases and self-satisfaction at being able to figure out how to buy some one else's used car sitting half a world away. They described the Internet as allowing them to participate within a global economy and to locate cars that would get good gas mileage and run forever. Imagining the Internet as a global marketplace has a different resonance in a rural community with limited access to commodity exchange, particularly to major appliances and automobiles. Machans shared excited accounts of the movement of this object that was manufactured in Japan and driven thousands of kilometers there, sold online, put in a cargo container, shipped across the seas, and delivered by truck to a person in Macha who could drive it for thousands of kilometers, evoking Arjun Appadurai's work on the social life of objects.[27] One car owner estimated that about seven thousand people in the area have bought used Japanese cars online.[28]

Within such conditions the resource economy of the Internet comes full circle. The movement of water generates electricity that energizes the movement of data, which catalyzes and completes an online transaction that results in the movement of a manufactured object from freeways in Japan, along shipping corridors on the high seas, through dirt roads in southern Zambia, and into the circuits of daily life in Macha. Internet access relies on hydropower at one end and intensifies local demand for gasoline on the other. Like electricity and water, gas supplies are limited in Macha. There is a makeshift gas station

Figure 5.8. This small gas station in Macha supports a growing number of used Japanese cars, which Machans have purchased online in recent years. Photo by author.

(see figure 5.8) in front of Macha's Blue Sky market that stays open until supplies run out, usually for a few hours. Gas is transported from Choma in large plastic tubs and sold on the side of the road in plastic bottles for 12.50 kwachas ($2.40) a liter. Thus as Internet access facilitates a sense of participation in the global economy, it alters local resource demands and modes of mobility such that humans, bicycles, oxen, dogs, and chickens now share the same dirt roads with Toyotas.

Never Heard of the Internet

While many of the women we spoke with carry water each day to sustain life, many had never heard of the Internet. Working with a translator, we talked to several groups of women at the fire camps outside the hospital where they prepare food and wait for family members receiving medical treatment. Most of these women, ages fifteen to seventy-five, live far beyond Macha's center, are part of extended families that live in traditional homesteads, and have had little, if any, formal education. When I asked if they had ever used or heard of the Internet, most simply shook their heads and seemed indifferent. And while

Figures 5.9a and 5.9b. Many women we interviewed at the fire camps near Macha's hospital indicated they have neither heard of nor used the Internet and expressed disinterest in it. Photos by author.

most had never heard of the Internet or used a computer, some had mobile phones. Several explained that they did not have enough money to buy talk time and used the mobile phone primarily to "ping" others and receive calls. They provide their numbers to family and friends so that they can receive incoming calls, but rarely can they afford to make outgoing calls. (Talk time can be purchased from local merchants as scratch cards in increments from .05 kwachas or thirty cents to fifty kwachas or ten dollars.) In addition to using the phone as a *receiver*, some prefer to use the mobile phone for group conversations rather than person-to-person phone calls. This way more people can participate in conversations without individually paying for talk time. On multiple occasions we witnessed people clustered around a single mobile phone with the speaker function activated, participating in a group conversation with a party on the other end, which suggests that Machans are using mobile telephony in ways that reconfigure and extend their oral cultures.

The discovery of how few Machan women knew about computers or the Internet generated discussion among members of our research team. On the one hand, the women's responses represented a gendering of the global digital divide, suggesting that rural Zambian women are geographically, socially, and economically positioned in ways that inhibit their capacity to learn computing and access the Internet, even though their labor and daily living routines (carrying water, farming, preparing food) may support others' Internet use. On the other hand, whether Machan women should or should not know about or use the Internet may ultimately be a question for them to broach on their own terms and decide. The women we spoke with seemed somewhat disinterested in the question and relatively content without the technology, foregrounding the reality that the digital divide may be as much an invention of Western humanitarianism and/or digital capitalism as it is a salient concern among Macha's rural residents. That so many women have never heard of the Internet caused me to reassess the very purpose of our project and to question whether or not we should be in Macha at all, particularly since the Dutch-supported ICT initiative had fallen apart and caused conflict in the community that led to a federal lawsuit in Zambia. Since there is no way for Westerners to engage in collaborative ICT work without the baggage of colonial pasts, development ideologies, and class and power hierarchies, and since we inherit and, in some cases, unwittingly evoke or reenact these conditions, how can international research collaboration be organized to craft imaginings and uses of ICTs that will expose, recalibrate, and reorder such relations? In the context of transnational feminist politics, is it vital for Machan women to know about and use the Internet? Certainly, there are many ways in which Internet access could support Machan women, but under what conditions should the introduction of ICTs within their lives take place? How can ICTD research be used to stage interactions geared toward the introduction of technological potentials and possibilities as opposed to idly advancing deterministic agendas of technological integration, adoption, and revelry?

Though these questions persist, our Machan partners have told me that our work together has been useful because it provided an occasion to talk about and reflect upon the ways ICT use has been reconfiguring everyday life, movement, religion, education, commerce, and social relations in their community, something most of them had not discussed before. In the meantime, there has been much turmoil and uncertainty in relation to Internet service in the community. After LinkNet went bankrupt in 2012, its two leaders—one Dutch and one Machan—received death threats, and the Dutch leader left the community permanently. Remaining LinkNet staff worked for months without compensation to try and sustain the community's Internet access. Eventually they had to leave the area to look for work, and the mission seized all of LinkNet's equipment, much of which now sits dormant and is off limits to anyone in the community. The mission's leadership also cut electrical power to the water tower, which shut down the functionality of all antennas, including that belonging to the community radio station, for more than a month. Most Machans who used the Internet between 2012 and 2013 were doing so through their mobile-phone service provider. Figuring out how to proceed with our research in the midst of these conditions has been tricky, but our collaboration continues.

Conclusion

This fieldwork in Macha fundamentally altered the manner in which I imagine "Internet infrastructure" and its materiality. Site visits, interviews with Machans, technical installations and tests, and everyday experiences in the community brought forth the complexity of the Internet's operational dynamism—its contingency upon the coordinated appropriation of natural resources, electricity and batteries, and human biopower. In order for the Internet to become a widely accessible and useable "media infrastructure" in Macha, it will be necessary for Machans to collectively determine whether that is what *they* want. This will involve deciding whether to organize the community's limited resources to strengthen the local link to the national power grid or invest in reliable off-the-grid alternatives. It will involve acquiring computers, mobile devices, and software and creating educational programs to support digital literacy. And it will involve fostering local interest in Internet use beyond acts of downloading information from elsewhere, and utilizing this infrastructure to support and reinvent Machan knowledge practices and ways of life.

Notes

This research is related to a four-year interdisciplinary study, "VillageNet: Intelligent Wireless Networks for Rural Areas," funded by the National Science Foundation, 2011–2015. I thank Chief Macha, Consider Mudenda, Gertjan van Stam, Elton Munguya, Fred Mweetwa, Abby Hinsman, Lindsay Palmer, Mariya Zheleva, David Johnson, Elizabeth Belding, and the people in the community of Macha for helping with this research. I thank Silbalwa Ntabeni for providing translation services during our fieldwork.

1. For purposes of this chapter, "dollars" refer to U.S. currency.

2. For a discussion of these issues see Ineke Buskens, "Agency and Reflexivity in ICT4D Research: Questioning Women's Options, Poverty, and Human Development," *Information Technologies and International Development* 6, special edition (2010): 19–24; Alison Gillwald, "The Poverty of ICT Policy, Research, and Practice in Africa," *Information Technologies and International Development* 6, special edition (2010): 79–88; A. Glenn Maail, "User Participation and the Success of Development of ICT4D Project: A Critical Review," available at http://www.globdev.org/files/Shanghai%20Proceedings/ 5%20REVISED%20Maail_UserParticipationICT4DSuccess.pdf (accessed January 16, 2014); Andrew Skuse and Thomas Cousins, "Getting Connected: The Social Dynamics of Urban Telecommunications Access and Use in Khayelitsha, Cape Town," *New Media and Society* 10, no. 1 (2008): 9–26.

3. O3B has also become the brand name of a new satellite venture to deliver Internet services in emerging markets. See the company's website, http://www.o3bnetworks .com.

4. For a discussion of the second-hand computers in Ghana and Namibia, respectively, see Jenna Burrell *Invisible Users: Youth in the Internet Cafés of Urban Ghana* (Cambridge, Mass.: MIT Press, 2012); and Steven J. Jackson, Alex Pompe, Gabriel Krieshok, "Repair Worlds: Maintenance, Repair, and ICT for Development in Rural Namibia," *CSCW' 12*, February 11–15, 2012, Seattle, Washington.

5. Conference call with Gertjan van Stam, David Johnson, and Dick Uyttewaal, Macha, Zambia, July 5, 2012.

6. For more detail on the history of Macha Works see Gertjan van Stam, "Placemark: Macha" (2011); Gertjan van Stam and G. van Oortmerssen, G., "Macha Works!" *Frontiers of Society On-Line*, 2010; Raleigh J. Bets, G. van Stam, and A. Voorhoeve, "Modeling and Practise of Integral Development in Rural Zambia: Case Macha," in *Africomm 2012*; and Gertjan van Stam, "Information and Knowledge Transfer in the Rural Community of Macha, Zambia," *Journal of Community Informatics* 9, no. 1 (2013).

7. Lisa Parks, "Satellite Footprints: Imparja TV and Postcolonial Flows in Australia," in *Cultures in Orbit: Satellites and the Televisual* (Durham, N.C.: Duke University

Press, 2005), 47–76; Lisa Parks, "Walking Phone Workers," in *The Routledge Handbook of Mobilities*, ed. Peter Adey et al. (London: Routledge, 2013), 243–55.

8. Lisa Parks, "Earth Observation and Signal Territories: Studying U.S. Broadcast Infrastructure through Historical Network Maps, Google Earth, and Fieldwork," *Canadian Journal of Communication* 38 (2013): 1–24; "Postwar Footprints: Satellite and Wireless Stories in Slovenia and Croatia," in *B-Zone: Becoming Europe and Beyond*, ed. Anselm Franke (Barcelona: ACTAR, 2007), 306–47; "Where the Cable Ends: Television in Fringe Areas," in *Cable Visions: Television Beyond Broadcasting*, ed. Sarah Banet-Weiser, Cynthia Chris, and Anthony Freitas (New York: New York University Press, 2007), 103–26.

9. Eric Michaels, *Bad Aboriginal Art* (Minneapolis: University of Minnesota Press); Faye Ginsburg, "Embedded Aesthetics: Creating a Discursive Space for Indigenous Media," *Cultural Anthropology* 9, no. 2 (1993): 365–82; Heather Horst and Daniel Miller, *The Cell Phone: An Anthropology of Communication* (New York: Bloomsbury Academic, 2006); Brian Larkin, *Signal and Noise* (Durham, N.C.: Duke University Press, 2009); Jenna Burrell, *Invisible Users* (Cambridge, Mass: MIT Press, 2012).

10. Diana Coole and Samantha Frost, eds., *New Materialisms: Ontology, Agency, and Politics* (Durham, N.C.: Duke University Press, 2012), 9–10.

11. See James Ferguson, *Expectations of Modernity: Myths and Meanings of Urban Life on the Zambian Copperbelt* (Berkeley: University of California Press, 1999).

12. See Dean Mulozi, "Rural Access: Options and Challenges for Connectivity and Energy in Zambia," eBrain Forum of Zambia and IICD, January 2008, 24, available at http://www.iicd.org/articles/rural-access-to-the-internet-in-zambia-options-and-challenges#sthash.7B1vth5g.dpuf (accessed January 16, 2014).

13. Also see, Consider Mudenda, David Johnson, Lisa Parks, and Gertjan van Stam, AFRICOMM 2013: Fifth Annual IEEE EAI Conference on e-Infrastructure and e-Services for Developing Countries, Blantyre, Malawi, November 25–28, 2013.

14. Tiziana Terranova, *Network Culture: Politics for the Information Age* (London: Pluto, 2004), 47.

15. Rick Maxwell and Toby Miller, *Greening the Media* (Oxford: Oxford University Press, 2012), 29–30.

16. Mr. Mugonke, interview with research partners Peter Miyanda and Trywell Maliko, July 3, 2012.

17. Steven J. Jackson, "Rethinking Repair," in *Media Technologies: Essays on Communication, Materiality and Society*, ed. Tarleton Gillespie et al. (Cambridge, Mass.: MIT Press, 2013), 222.

18. See A. Anand, V. Pejovic, E. Belding, and D. L. Johnson. "VillageCell: Cost-Effective Cellular Connectivity in Rural Areas," ICTD '12, Atlanta, Georgia, March 2012; and M. Zheleva, A. Paul, D. L. Johnson, and E. Belding, "Kwiizya: Local Cellular Network Services in Remote Areas," ACM MobiSys, Taipei, Taiwan, July 2013.

19. David L. Johnson, Veljko Pejovic, Elizabeth M. Belding, and Gertjan van Stam, *ACM DEV Conference, ACM, 2012, 7.*

20. Ibid.

21. Dean Mulozi, "Rural Access: Options and Challenges for Connectivity and Energy in Zambia," eBrain Forum of Zambia and IICD, January 2008, 24, available at http://www.iicd.org/articles/rural-access-to-the-internet-in-zambia-options-and-challenges#sthash.7B1vth5g.dpuf (accessed January 16, 2014).

22. "Landlocked Countries Pin Their Hopes on EASSy Cable," *The East African*, undated, available at http://www.theeastafrican.co.ke/business/-/2560/679872/-/view/printVersion/-/sc69iq/-/index.html (accessed January 16, 2014).

23. We worked with the following partners in 2012: Consider Mudenda, Trywell Maliko, Cashmore Sikabanze, Peter Miyanda, Ruth Chilweza, Nina Kyalifungwa, Angela Kafute, Sibalwa Ntabeni, Consider's wife, Shadrek Llumuno, Ascent Milimo, Austin Sinzala; and the following partners in 2013: Consider Mudenda, Gracious Chizanda, Mutinta Maambo, Evis Muunga, Peter Miyanda, Bernard Sishumba, Calvin Muunta, Shadrek Llumuno.

24. Lisa Parks and Lindsay Palmer, "Critical Reflections on Rural ICTD Research: Collaborative Ethnography in Rural Zambia," work in progress.

25. Johnson, *ACM DEV*.

26. Gertjan van Stam and Fred Mweetwa, "Community Radio Provides Elderly a Platform to Have Their Voices Heard in Rural Macha, Zambia," *Journal of Community Informatics* 8, no. 1 (2012), available at http://ci-journal.net/index.php/ciej/article/view/870/832 (accessed January 16, 2014).

27. Arjun Appadurai, *The Social Life of Things: Commodities in Cultural Perspective* (Cambridge: Cambridge University Press, 1988).

28. Email from Consider Mudenda to Lisa Parks, March 18, 2013.

CHAPTER 6

The Art of Waste

Contemporary Culture and Unsustainable Energy Use

TOBY MILLER

The fundamental message of this chapter is that contemporary culture hinges upon unsustainable energy use. Whether the topic is fine art or reality TV, each one is complicit with our global environmental crisis. This development also articulates to a new form of diminished worker power—the cognitariat. Together, these tendencies present artists with serious ethical, political, and economic questions. Many of them are responding to those challenges in constructive, reflective ways that can stimulate the rest of us to join the dots and appreciate just how dangerous digital culture is to our world, even as we rely so much on it.

A fascinating, unholy, productive convergence is underway: while artists are becoming more connected to the global communications infrastructure due to their digital delights, workers in that infrastructure are shifting, like so many others, toward the contingent, discounted labor force that artists have known and occupied for decades. Electronic or e-waste artists operate in a sector that relies on discounted labor and hence exemplifies wider work trends, even as their art incarnates a vanguard ecological awareness. The particular focus of the chapter is, therefore, on the art of e-waste and the question of artistic labor. I hope that readers who produce or enjoy all forms of media culture will think anew about their practices of work and consumption thanks to the provocations that e-waste artists offer, both industrially and textually.

My methods in this piece derive from political economy, environmentalism, and cultural studies, focused on the material relations of meaning and the interplay of cultural subjectivity, ecology, and power. This eco-materialism emphasizes the materiality of discourse and the discursivity of materiality. That is to say, it refuses the notion that objects lack meaning or meanings exist independent of objects. It is also profoundly connected to the fundamental question, *"Cui bono?"* when discussing the allocation, utilization, and impact of resources as they touch the lives of workers, citizens, and all the Earth's creatures.[1]

Some scenery needs to be in place to show how culture contributes to ecological problems and models postindustrial labor. Let's start by clearing our minds of cant: for all the recycling bins that we assiduously fill and empty, we live in an age of waste. A seemingly disposable world is inexorably disposing of itself. The 2013 report from the Intergovernmental Panel on Climate Change makes that plain.[2] The next step is a political-economic rather than a purely ecological one: to recognize that the ultimate side effect of rapacious capitalist growth will be, paradoxically, the *end* of rapacious capitalist growth. True believers' faith that the market is a self-limiting and self-sustaining jewel of human nature may well have the effect of *ending* human nature. So good luck with that one.

We were all brought up believing that mining and manufacturing were the world's principal polluting culprits. The difficult news for media and cultural studies and the art world is that our beloved electronics are also crucial components of this destructiveness. Their toxic parts, forms, and norms pervade our world, from old fat-screen television sets to modish computing clouds, from museums' carbon footprints to Facebook and Twitter engagement announcements. The deleterious effects of these technologies is felt in the mines and factories that produce them, the offices and cars that house them, and the municipal dumps and fire pits that bury them.[3]

Yet such gadgets and sites are frequently regarded as signs of transcendent progress in a credulous world where life is routinely reinvented as an unconscious palimpsest of the past, driven by institutionalized amnesia.[4] This compulsive repetition of a seemingly unfamiliar history is nowhere clearer than in techno-futurism's predictions of social change. Seventy years ago, George Orwell described technologically determinist fantasies in words that resonate today with the same arid irony that first animated them:

> Reading recently a batch of rather shallowly optimistic "progressive" books, I was struck by the automatic way in which people go on repeating certain phrases which were fashionable before 1914. Two great favourites are "the abolition of distance" and "the disappearance of frontiers." I do not know how often I have

met with the statements that "the aeroplane and the radio have abolished distance" and "all parts of the world are now interdependent."[5]

Today's *mantra* is very similar to the fantasy that Orwell noticed and abjured all those years ago: utopian yearnings for a world free of institutional constraints. The latest media technologies are said to obliterate geography, sovereignty, and hierarchy in an alchemy of truth and beauty. A deregulated, individuated, technologized world makes consumers into producers, frees the disabled from confinement, encourages new subjectivities, rewards intellect and competitiveness, links people across cultures, and allows billions of flowers to bloom in a postpolitical cornucopia. It is a bizarre utopia. People fish, film, fornicate, and finance from morning to midnight. Consumption is privileged, production is discounted, and labor is forgotten. Powerful communications institutions cleave to themselves a sense of universal enlightenment through the wires and wireless that their products offer individuals. So Facebook features "Peace on Facebook" and claims the capacity to "decrease world conflict" through intercultural communication, while Twitter modestly announces itself to be "a triumph of humanity."[6] Machinery, not political-economic activity, is the guiding light: technology and consumption rather than activism and citizenship.

The wonderfully named Progress and Freedom Foundation's *Magna Carta for the Information Age*, for instance, proposes that political-economic gains made for democracy since the thirteenth century have been eclipsed by technological ones:

> The central event of the 20th century is the overthrow of matter. In technology, economics, and the politics of nations, wealth—in the form of physical resources—has been losing value and significance. The powers of mind are everywhere ascendant over the brute force of things.[7]

The foundation has closed its doors, no doubt overtaken by progress, but its ahistorical Whiggish discourse of an inevitably unfurling liberty for all continues to ring loudly in our ears, tinnitus-like. *Time* magazine exemplified this love of a seemingly immaterial world when it chose "You" as 2006's "Person of the Year," because "You control the Information Age. Welcome to your world."[8] On the liberal left, the *Guardian* is prey to the same touching warlockcraft: someone called "You" heads its 2013 list of the hundred most important folks in the media.[9] Rupert Murdoch was well behind, at number eight. You, Rupert, head to head. No contest, really.

To illustrate the pervasiveness of this magic via academic/policy examples, consider these three cases of barely contained scholarly and media exultation. First, bourgeois economists claim that cell phones have streamlined markets in the Global South, enriching people in zones where banking, economic in-formation, and market data are scarce. Fantastic claims are made for the mar-vel of mobile telephony in places that lack electricity, plumbing, fresh water, hospital care, and the like. These include "the complete elimination of waste" and massive reductions in poverty and corruption through the empowerment of individuals.[10] *Forbes* magazine and the International Monetary Fund lap this type of research up, deeming it "seminal"—as they would.[11] Nielsen, the world's leading media ratings company, published an unimaginably crass paean that began, "Africa is in the midst of a technological revolution, and nothing illus-trates that fact [more] than the proliferation of mobile phones," before casually noting that "more Africans have access to mobile phones than to clean drinking water."[12]

Second, the world seems agog these days in the face of three-dimensional printers, which promise the cheap and spectacular production of art, among many other applications. But while some analysts predict that 3-D printers will have positive ecological effects by reducing the carbon used to transport goods,[13] many use heated thermoplastic extrusion and deposition. Numerous factory studies have associated such processes with dangerous aerosol emis-sions, but minimal investigation has been done into the new printers, which generally lack exhaust ventilation or filtration systems. The first published study looked at ultrafine particle (UFP) production. It found that UFP emissions in an office using 3-D printing were alarmingly sizeable. Why alarming? UFPs can easily deposit themselves in people's lungs, airways, and brains, producing high concentrations of adsorbed, absorbed, and condensed compounds. The epidemiological record corresponds to cardiorespiratory mortality, strokes, and asthma.[14] E-waste issues galore of this type arise with new electronic textile art forms that merrily discard electronics en route to greater cultural glory.[15] Translating that research into the cultural world can improve public health and stimulate a healthy skepticism about techno-rhetoric.

Third, recall the publicity generated when Kelvin Doe/DJ Focus, a fifteen-year-old Sierra Leonean, was invited to MIT in 2012 because he had constructed a radio station from detritus in trashcans. More than two million online view-ings of the university's video about him in just one week testify to the appeal of this apparently unlikely story of a prodigy from the Global South who was constructed as embodying the need to replace aid programs with individual initiative. *Fast Company* included him in its list of "100 Most Creative People in Business 2013" under the soubriquet "The Philanthropic Prodigy."[16]

That account ignored an alternative one. It might have analyzed his achieve-ment as an impressive moment in centuries of skillful cultural ragpicking, a heritage that illustrates the constitutive power of creativity and collectivity as well as colonialism and pollution in forging conditions of existence for the young entrepreneur.[17] Stories like his can draw us into the materiality and inequality at the heart of media technologies, and question their utility—but only if these versions are critical and sidestep contemporary fashion.

Technocentric utopianism is an extended dalliance with consumer communication technology's supposedly innate capacity to endow its users with transcendence. It shies away from addressing unequal infrastructural and cultural exchange.[18] The discourse buys into individualistic fantasies of reader, audience, consumer, and player autonomy—the neoliberal intellectual's wet dream of music, movies, television, art, literature, performance, and everything else converging under the sign of empowered, creative audiences.

The New Right of cultural and communication studies invests with unparalleled gusto in this dream, which is populated by Schumpeterian entrepreneurialism, evolutionary economics, and creative industries. It has never seen an "app" it did not like or a socialist idea it did. Faith in devolved culture-making amounts to a secular religion, offering transcendence in the here and now via a "literature of the eighth day, the day after Genesis."[19] Hence the Australian Council for the Humanities, Arts and Social Sciences informing the country's Productivity Commission that this is a "post-smokestack era"[20]—a blessed world for workers, consumers, and residents, with residues of code rather than carbon. An astonishing claim from a country that survives on dirty-power exports that make it per capita among the greatest polluters in history[21]—yet why spoil a good story?

But as Orwell realized, the story is more complex. Max Weber insisted that technology was principally a "mode of processing material goods,"[22] and Harvey Sacks emphasized "the failures of technocratic dreams[:] that if only we introduced some fantastic new communication machine the world will be transformed."[23] The Political Economy Research Institute's 2013 *Misfortune 100: Top Corporate Air Polluters in the United States* placed half a dozen media owners in the first fifty.[24] Cultural production relies on the exorbitant water use of computer technology, while making semiconductors requires hazardous chemicals, including carcinogens. At current levels, residential energy use of electronic equipment will rise to 30 percent of the overall global demand for power by 2022, and to 45 percent by 2030, thanks to server farms, data centers and the increasing time people around the world spend watching and adding to screens.[25] So rather than seeing new communications technologies as magical agents that can produce market equilibrium and hence individual and collective happiness, we should note their other effects—and their continued exclusivity. In 2011, the cost of broadband in the Global South was 40.3 percent of average individual Gross National Income (GNI). Across the Global North, by comparison, the price was less than 5 percent of GNI per capita.[26]

E-Waste

Away from questions of content and use, when old and obsolete cell phones or other communication technologies are junked, they become electronic waste, the fastest-growing component of municipal cleanups around the Global North. E-waste has generated serious threats to worker health and safety wherever plastics and wires are burnt, monitors smashed and dismantled, and circuit boards grilled or leached with acid, while the toxic chemicals and heavy metals that flow from such practices have perilous implications for local and downstream residents, soil, and water. The accumulation of electronic hardware causes grave environmental and health harm as noxious chemicals, gases, and metals from wealthy nations seep into landfills and water sources across Malaysia, Brazil, South Korea, China, Mexico, Vietnam, Nigeria, and India, inter alia. The e-waste ends up there after export and import by "recyclers" who eschew landfills and labor in the Global North in order to avoid the higher costs and regulatory oversight of recycling in countries that prohibit such destruction to the environment and workers. Businesses that forbid dumping in local landfills as part of their corporate policies merrily mail it elsewhere. In that "elsewhere," preteen girls pick away without protection at discarded televisions and computers, looking for precious metals to sell—less romantic ragpickers than MIT's Kelvin Doe.[27]

This material reality remains invisible to the new-media clerisy and bourgeois economics alike, but it *has* been recognized in the technocratic cloisters of communications diplomacy. In keeping with prevailing shibboleths, the International Telecommunication Union (ITU) predicts that communications technologies will connect the 6.5 billion residents of the Earth by 2015, enabling everyone to "access information, create information, use information and share information." This development "will take the world out of financial crisis," principally thanks to developing markets.[28] But the ITU also acknowledges that communications technologies cause grave environmental problems, so it presses for "climate neutrality" and greater efficiency in energy use.[29] The 2008 World Telecommunication Standardization Assembly in South Africa encouraged members to reduce the carbon footprint of communications, in accord with the United Nations Framework Convention on Climate Change.[30]

In a similar vein, the Organization for Economic Co-operation and Development (OECD) says communications can play a pivotal role in developing service-based, low-polluting economies in the Global South through energy efficiency, adaptation to climate change, mitigation of diminished biodiversity, and diminished pollution. But it cautions that such technological advances can produce negative outcomes. For example, remote sensing of marine life may encourage unsustainable fishing.[31]

Then there is that delightful metaphor we are all now using: "the cloud." It signifies the place where all good software goes for rest and recuperation, emerging on demand, refreshed and ready to spring into action. Seemingly ephemeral and natural—benign necessities of life, clouds rain then go away—nothing could be further from the truth when it comes to the power-famished server farms and data centers rendered innocent by this perverse figure of speech.

The U.S. National Mining Association and the American Coalition for Clean Coal Electricity gleefully avow that the "Cloud Begins with Coal."[32] They boast that the world's information and communications technologies use fifteen hundred terawatt hours each year—equivalent to Japan and Germany's overall energy use combined. That's 10 percent of global electricity—and 50 percent more than aviation. The association and the coalition even quote Greenpeace,[33] against the grain, on the horrendous environmental implications of data centers, as support for the endless coal opportunities to come! Big mining and big coal just can't help themselves, so excited are they by the importance of their polluting ways for the present and future of the cloud. Meanwhile, Google disclosed in 2011 that its annual carbon footprint was almost equal to that of Laos or the United Nations Organization, largely due to running its search engines through clouds.[34]

The Cognitariat

What about the making of culture—the things that reside in the cloud? Aren't corporate and governmental cultural gatekeepers and hegemons fundamentally undermined by the new technological possibilities of creation and distribution, which can scarcely be likened to the horrors of mining and manufacturing in their impact on either work or the environment? In the new era, readers become writers, listeners transform into speakers, viewers emerge as stars, fans are academics, and vice versa. The economy glides into an ever-greener postindustrialism. The comparatively cheap and easy access to making and circulating meaning afforded by Internet media and genres is thought to have eroded the one-way hold on culture that saw a small segment of the world as producers and the larger segment as consumers, even as it makes for a cleaner economy. New technologies supposedly allow us all to become simultaneously cultural consumers and producers (prosumers)—no more factory conditions, no more factory emissions. More artists, and more power to artists.[35]

In this era of the "prosumer," anyone can *be* an artist. Zine writers are screen-writers. Bloggers are copywriters. Children are columnists. Bus riders are jour-nalists. And think of the job prospects that follow! Coca-Cola hires African Americans to drive through the inner city selling soda and playing hip-hop. AT&T pays San Francisco buskers to mention the company in their songs. Urban performance poets rhyme about Nissan cars for cash, simultaneously hawking, entertaining, and researching. Subway's sandwich commercials are marketed as made by teenagers. Cultural-studies majors become designers. Graduate students in New York and Los Angeles read scripts for producers and then pronounce on whether they tap into audience interests. Precariously employed part-timers spy on fellow-spectators in theaters to see how they respond to coming attractions. Opportunities to vote in the Eurovision Song Contest or a reality program disclose the profiles and practices of viewers, who can be monitored and wooed in the future. End-user licensing agreements en-sure that players of corporate games online sign over their cultural moves and perspectives to the very companies they are paying in order to participate.[36]

In other words, corporations are using discounted labor. Business leeches want flexibility in the people they employ, the technologies they use, the places where they do business, and the amounts they pay—and inflexibility of owner-ship and control. The neoclassical doxa preached by neoliberal chorines favor an economy where competition and opportunity cost are in the litany and dis-sent is unforgiveable, as crazed as collective industrial organization. In short, "decent and meaningful work opportunities are reducing at a phenomenal pace in the sense that, for a high proportion of low- and middle-skilled workers, full-time, lifelong employment is unlikely."[37] Even reactionary bodies like the U.S. National Governors Association recognize the reality: "Routine tasks that once characterized middle class work have either been eliminated by technological change or are now conducted by low-wage but highly skilled workers."[38] Cul-tural workers, from jazz musicians to street artists, have long labored without regular compensation and security. That models the expectations we are all supposed to have today, rather than our parents' or grandparents' assumptions about life-long, or at least steady, employment.

Hence the success of Mindworks Global Media, a company outside New Delhi that provides Indian-based journalists and copyeditors who work long-distance for newspapers whose reporters are supposedly in the United States

and Europe. There are 35 percent to 40 percent cost savings.[39] Or consider the
advertising agency Poptent, which undercuts big competitors in sales to major
clients by exploiting prosumers artists' labor in the name of "empowerment."
That empowerment takes the following form: Poptent pays the creators of
homemade commercials seventy-five hundred dollars; it receives a management fee of forty-thousand dollars; and the buyer saves about three hundred
thousand dollars on the usual price.[40] The slogan says it all:

> Accelerate your video career
> Access the biggest brands. Build your network. Get paid.

Antonio Negri redeployed the concept of the cognitariat from the Reaganite
futurist and digital *Magna Carta* signatory Alvin Toffler to account for this phenomenon.[41] Negri defines the cognitariat as people undertaking casualized
cultural work who have heady educational backgrounds yet live at the uncertain
interstices of capital, qualifications, and government in a post-Fordist era of
mass unemployment, limited-term work, and occupational insecurity. They
are sometimes complicit with these circumstances because their identities are
shrouded in autotelic modes of being: work is pleasure and vice versa; labor
becomes its own reward.[42] The art world is a model.

The wider culture industries largely remain controlled by media and communications conglomerates, which frequently seek to impose artist-like conditions
on their workforces (the cable versus broadcast TV labor process is a notorious
instance). They gobble up smaller companies that invent products and services, "recycling audio-visual cultural material created by the grassroots genius,
exploiting their intellectual property and generating a standardized business
sector that excludes, and even distorts, its very source of business," to quote *The
Hindu.*[43] In other words, the cognitariat—interns, volunteers, contestants, and
so on—creates "cool stuff" whose primary beneficiaries are corporations.[44]

At the same time, the governing assumption of Internet and arts boosters
is that culture is an endlessly growing resource that can dynamize both society
and economy. The Australian Academy for the Humanities calls for "research
in the humanities and creative arts" to be tax-exempt based on its contribution
to research and development, and subject to the same surveys of "employer
demand" as the professions and sciences as a quid pro quo.[45] The Australian
Research Council's Centre of Excellence for Creative Industries and Innovation
has solemnly announced an "industryfacing [sic] spin-off from the centre's
mapping work, Creative Business Benchmarker."[46] In partnership with the Arts
and Humanities Research Council, the UK's National Endowment for Science,
Technology and the Arts says, "The arts and humanities have a particularly

strong affiliation with the creative industries" and provide research that "helps to fuel" them, in turn boosting innovation more broadly.[47] True believers all; none of the issues raised in this chapter seem to touch them—unlike the ITU or OECD.

The Good News

So that all looks rather bleak, doesn't it? If you are a credulous cybertarian and you have kept reading so far, is there any good news? If you are that much simpler being, a skeptic, where is the joy? And no matter who you are—where is the art? It is in the title, but does it figure here, apart from the spread of exploitative labor practices into the core of an allegedly postindustrial economy?

Environmental art covers many works that directly and indirectly represent nature. Older examples, from the canon of high European culture, might be Claude Monet's *London Series* or John Constable's *Clouds*.[48] More tendentious instances from today include such nonrepresentational, performative pieces as Richard Long's *A Line Made by Walking*, James Turrell's *Skyspace*, or Olafur Eliasson's *The Weather Project*, which assume nature is occupied and shaped by humanity, and vice versa.[49]

They appreciate the lesson of Charles Babbage, the mythic founder of programmable computation. Almost two centuries ago, he noted the partial and ultimately limited ability of humanity to bend and control natural forces without unforeseen consequences: "The operations of man . . . are diminutive, but energetic during the short period of their existence: whilst those of nature, acting over vast spaces, and unlimited by time, are ever pursuing their silent and resistless career."[50] E-waste artists are alert to these questions, both in terms of their own practice and the wider world of cybertarianism. While keen to use mixed-media methods and new technologies, they understand full well the risk as much as the potential that cleaves to gadgetry.[51] Consider *Arte Povera*'s classic use of found materials, or such artists as Jessica Millman,[52] Miguel Rivera,[53] Sudhu Tewari,[54] Natalie Jeremijenko, Nome Edonna's deviant art,[55] Chris Jordan,[56] Erik Otto,[57] or Jane Kim.[58] In 2014, Chris Jordan built the world's biggest e-waste artwork in Australia: a huge cell phone entitled "23." Made by the artist with schoolchildren and eight thousand discarded phones, it stood for the 23 million unused cell phones sitting around Australian buildings, mute testimony to an insatiable culture of built-in obsolescence.[59]

Yona Friedman focuses on redeployment rather than originality,[60] while Julie Bargmann and Stacy Levy start with the creative cleanup of waste rather than concluding with a painstaking one.[61] The Carnegie Endowment's *Foreign*

Policy magazine circulated into the mainstream Natalie Behring's stunning collection of photos from "Inside the Digital Dump."[62] Amsterdam's Urban Screens electronic billboards encourage active citizenship in public spaces, as do Ars Electronica of Linz and Melbourne's Federation Square.[63] Yuri Suzuki uses e-waste to rematerialize the map of the London Underground, encouraging people to think of iconic representations like the Roundle and the Circle Line as perennial, thus inviting them to ponder the little black boxes in their lives, from phones to tablets, as potentially reusable rather than necessarily replaceable items.[64] Peter McFarlane draws on his sales experience hawking built-in obsolescence to criticize "innovation." He makes discarded circuit boards simulate fossil life—an ironic comment on the path to self-destruction with which we began.[65] Rodrigo Alonso turns electronic trash into designer furniture.[66] Mairo Cacedo Langer reboots robots as Robo Planters, wacky pot holders with personality.[67] ReFunct Media #5 is less concerned with end products than reimaging our relationship to the process of creating e-waste.[68] Dani Ploeger explores e-waste and feminist struggles in performance pieces such as "Waste Circuits" and through anal electrodes.[69] These works remind us of the materiality of e-waste in phenomenological terms, as does Beijing's 798 Art Zone and its reuse of e-waste.[70]

I want to focus here on work by Natalie Jeremijenko, who installed a Model Urban Development on the roof of the Postmasters Gallery's former headquarters in Chelsea, New York. The project featured seven residential housing developments, a concert hall, and other public amenities, powered by human food waste. The installation toyed with new conceptions of urban futures, reimagining our relationship to nonhuman organisms. Jeremijenko's work is referred to as experimental design, or xDesign, and explores the opportunity that new technologies present for progressive, pacific change.

One of Jeremijenko's renowned projects is "Feral Robotic Dogs," which finds her adapting fallen (or are they risen?) toys to sniff out environmental toxins (see figure 6.1). She hands them to victims of environmental racism, assisting them to identify and intervene in their situation.[71] Her description of "Feral Robotic Dogs" emphasizes several aspects of the project: fun—the joy of learning about robotic dogs; safety—the need to use machines to counter environmental racism; access—the importance of working alongside people traditionally excluded from the use of such gadgets; and recycling—the lesson heeded by so many great artists and designers: that there is value in tinkering with success as opposed to seeking newness. The best innovation builds upon rather than displaces what went before. Dogs created as asinine executive toys are recreated as activist art works.[72] One thinks here of Francis Alÿs, who makes "collector"

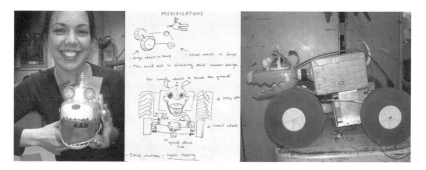

Figure 6.1. Feral robotic dogs, Elise (left and center) and Gollum (right). Elise was designed to seek out the toxin PERC from the ruins of the "American Linen" mass-quantity laundry business. Images courtesy of Natalie Jeremijenko.

toys from discarded magnets, cans, and other detritus to pick urban scraps while walking.[73]

How Stuff Is Made (HSIM, http://howstuffismade.org) is Jeremijenko's visual encyclopedia. It documents the physical processes, labor conditions, and environmental impact of contemporary manufacturing. Design and engineering students produce summative photo essays that describe these conditions of creation. HSIM reconsiders the responsibilities and capacities of design and engineering education in the light of sociopolitical constraints, organizational innovations, and globalization.[74]

What can such works of art do in broader political terms? This question has exercised thinkers of every epoch and kind, from Plutarch to Trotsky. When we ponder avant-garde uses of spectacle, it's easy to fall into either a critical camp or a celebratory one. The critical camp would say that rationality must be appealed to in discussions of climate change, and a progressive goal of capturing popular emotion will ultimately fail. Why? The silent majority doesn't like the avant-garde, marketing outspends art, such occasions preach to a light-skinned, middle-class eco-choir, media coverage is slender, and crucial decisions are made in golf carts, not galleries. Conversely, the celebratory camp would argue that a Cartesian distinction between hearts and minds is not sustainable, a sense of humor is crucial in order to avoid the image of environmentalists as finger-wagging scolds, corporate capital must be opposed in public, the media's need for vibrant textuality can be twinned with serious discussion as a means of involving people who are not conventional activists, and a wave of anti-élite sentiment is cresting. In 2013 *The Economist* predicted that the "silicon elite will cease to be regarded as geeks who happen to be filthy rich and become filthy rich people who happen to be geeks,"[75] with a backlash against cybertarianism inspired by Occupy[76] and Anonymous.[77]

Absent external evaluation of the social composition of art world participants, the nature of its coverage by old, middle-aged, and new media, and subsequent shifts in public opinion and reactions from lawmakers, it is difficult to be sure about the impact of such art. I generally incline toward the skeptic's view—but not in these instances. Why? Because I think the lugubrious hyperrationality often associated with environmentalism needs leavening through sophisticated, entertaining, participatory spectacle. A blend of dark irony, sarcasm, and cartoonish stereotypes can be twinned with a radical departure from a cultural world that lines up to exploit the Earth with gullible consumers pressing their noses to the Windows and Apples of this world, looking for operating systems as if they were upgrading their own bodies. Jeremijenko's work, inter alia, instantiates just such endeavor.

Conclusion

Simmel argued that:

> When we designate a part of reality as nature, we mean one of two things . . . an inner quality marking it off from art and artifice, from something intellectual or historical. Or . . . a representation and symbol of that wholeness of Being whose flux is audible within them.[78]

The oeuvre mentioned briefly above helps one imagine the relationship of a sustainable, democratic, and pleasurable life—a healthy Earth, a functioning global democracy, and fun—to art. E-waste artists translate scientific and activist ideas and found or invented materials, encouraging us to think of the imminent, not just the past and present.[79] This engages popular culture in an avant-garde way that can feed back into the everyday and in turn be made sense of by public-interest intermediaries as well as opportunistic commerce.

Perhaps I am describing/endorsing a very conventional view of art, such that it trials new forms of life that may be taken up by the mainstream. But I am seeking a different inflection, focused on the capacity of these works to exemplify and criticize a human and ecological disaster that must not be allowed to continue. Artists are uniquely placed to enliven such conversations, due to the centrality of their labor process to the spread and development of a cognitariat and their self-critical complicity with the environmental peril that is enabled by digital culture. Their creative reuse of waste as art challenges our upgrade society's culture of built-in obsolescence, while the curating of such work by museums can be part of a wider commitment against e-waste.[80]

William Morris asked some powerful questions a century and more ago about the links between labor, art, and the environment. He called for the art world to recognize its links to everyday life and problematize a Romantic separation of work from creativity:

> Of all the things that is likely to give us back popular art in England, the cleaning of England is the first and most necessary. Those who are to make beautiful things must live in a beautiful place. Some people may be inclined to say . . . that the very opposition between the serenity and purity of art and the turmoil and squalor of a great modern city stimulates the invention of artists, and produces special life in the art of today. . . . It seems to me that at best it but stimulates the feverish and dreamy qualities that throw some artists out of the general sympathy . . . men who are stuffed with memories of more romantic days and pleasanter lands, and it is on these memories they live.[81]

E-waste artists at their best inhabit a world where these antinomies are put into dialectical play. They use the freedom of art to demand secure labor and a sustainable environment. E-waste turns a post-smokestacks world of impermanent cultural employment upside down, making us rethink the ecological and employment dualities of the contemporary moment.

Notes

1. The chapter draws on previous work and extends this into the response of artists to related problems. Richard Maxwell and Toby Miller, *Greening the Media* (New York: Oxford University Press, 2012); and http://www.psychologytoday.com/blog/greening-the-media (accessed March 13, 2014).

2. Intergovernmental Panel on Climate Change, World Meteorological Organization, http://www.ipcc.ch/report/ar5/wg1/#.UkXnuGSG3UM (accessed March 13, 2014).

3. Lisa Parks, "Energy-Media Vignettes," *Flow* 19, no. 8 (2014), available at http://flowtv.org/2014/03/energy-media-vignettes/ (accessed March 13, 2014).

4. Herbert J. Gans, "Sociological Amnesia: The Noncumulation of Normal Social Science," *Sociological Forum* 7, no. 4 (1992): 701–10.

5. George Orwell, "As I Please," *Tribune*, May 12, 1944, available at http://telelib.com/authors/O/OrwellGeorge/essay/tribune/AsIPlease19440512.html (accessed March 13, 2014).

6. "A Cyber-House Divided," *Economist*, September 4, 2010, 61–62.

7. Esther Dyson, George Gilder, George Keyworth, and Alvin Toffler, *Cyberspace and the American Dream: A Magna Carta for the Knowledge Age*, version 1.2, Progress and Free-

dom Foundation, 1994, available at http://www.pff.org/issues-pubs/futureinsights/fi1.2magnacarta.html (accessed March 13, 2014).

8. Lev Grossman, "*Time*'s Person of the Year: You," *Time*, December 13, 2006, available at http://content.time.com/time/magazine/article/0,9171,1570810,00.html (accessed March 13, 2014).

9. "Media Guardian 100," *Guardian*, September 2, 2013, available at http://www.theguardian.com/media/series/mediaguardian-100-2013-1-100 (accessed March 13, 2014).

10. Robert Jensen, "The Digital Provide: Information Technology, Market Performance, and Welfare in the South Indian Fisheries Sector," *Quarterly Journal of Economics* 122, no. 3 (2007): 879–924.

11. Dean Karlan, "Every Which Way We Can," *Finance and Development* 49, no. 4 (2012), available at https://www.imf.org/external/pubs/ft/fandd/2012/12/karlan.htm (accessed March 13, 2014).

12. Jan Hutton, "Mobile Phones Dominate in South Africa," *Nielsen Wire*, September 30, 2011, available at http://blog.nielsen.com/nielsenwire/global/mobile-phones-dominate-in-south-africa (accessed March 13, 2014).

13. Thomas Campbell, Christopher Williams, Olga Ivanova, and Banning Garrett, *Could 3D Printing Change the World? Technologies, Potential, and Implications of Additive Manufacturing*, Atlantic Council, 2011, available at http://www.acus.org/files/publication_pdfs/403/101711_ACUS_3DPrinting.PDF (accessed March 13, 2014).

14. Brent Stephens, Parham Azimi, Zeineb El Orch, and Tiffanie Ramos, "Ultrafine Particle Emissions from Desktop 3D Printers," *Atmospheric Environment* 79 (2013): 334–39.

15. Andreas R. Köhler, Lorenz M. Hilty, and Conny Bakker, "Prospective Impacts of Electronic Textiles on Recycling and Disposal," *Journal of Industrial Ecology* 15, no. 4 (2011): 496–511.

16. David Lieberman, "Deadline's YouTube Channel Rankings," *Deadline.com*, November 23, 2012, available at http://www.deadline.com/2012/11/deadline%E2%80%99s-youtube-channel-rankings-2/#utm_source=sailthru&utm_medium=email&utm_campaign=breakingnewsalert; Hayley Hudson, "Kelvin Doe, Self-Taught Engineering Whiz From Sierra Leone, Wows MIT," *Huffington Post*, November 19, 2012, available at http://www.huffingtonpost.com/2012/11/19/kelvin-doe-self-taught-en_n_2159735.html; Jeff Chu, "The Philanthropic Prodigy," *Fast Company*, May 13, 2013, available at http://www.fastcompany.com/3009225/most-creative-people-2013/43-kelvin-doe (all accessed March 13, 2014).

17. Martin Medina, *The World's Scavengers: Salvaging for Sustainable Consumption and Production* (Lanham, Md.: AltaMira, 2007).

18. Christine L. Ogan, Manaf Bashir, Lindita Camaj, Yunjuan Luo, Brian Gaddie, Rosemary Pennington, Sonia Rana, and Mohammed Salih, "Development Communication: The State of Research in an Era of ICTs and Globalization," *Gazette* 71, no. 8 (2009): 655–70.

19. James Carey, "Historical Pragmatism and the Internet," *New Media and Society* 7, no. 4 (2005): 443–55.

20. CHASS, *CHASS Submission: Productivity Commission Study on Science and Innovation,* 2006, available at http://www.chass.org.au/submissions/pdf/SUB20060807TG.pdf (accessed March 13, 2014).

21. Simon Lauder, "Australians the 'World's Worst Polluters,'" *World Today,* September 11, 2009, available at http://www.abc.net.au/news/2009-09-11/australians-the -worlds-worst-polluters/1425986 (accessed March 13, 2014).

22. Max Weber, "Remarks on Technology and Culture," trans. Beatrix Zumsteg and Thomas M. Kemple, ed. Thomas M. Kemple, *Theory, Culture and Society* 22, no. 4 (2005): 23–38.

23. Harvey Sacks, *Lectures on Conversation,* vols. 1 and 2, ed. Gail Jefferson (Malden, Mass.: Blackwell, 2005).

24. See http://www.peri.umass.edu/toxicair_current/ (accessed March 13, 2014).

25. Maxwell and Miller, *Greening the Media.* Some of the citations it includes are: Jad Mouawad and Kate Galbraith, "Plugged-In Age Feeds Hunger for Electricity," *New York Times,* September 20, 2009; International Energy Agency, *Gadgets and Giga-watts: Policies for Energy Efficient Electronics—Executive Summary* (Paris: Organization for Economic Cooperation and Development, 2009), 5, 21; Climate Group, *Smart2020: Enabling the Low Carbon Economy in the Information Age* (London: Global Sustainability Initiative, 2008), 8–23; Simon Hancock, "Iceland New Home of Server Farms?," *BBC News,* October 10, 2009, available at news.bbc.co.uk/go/pr/fr/-/2/hi/programmes/ click_online/8297237.stm (accessed March 13, 2014); Organisation for Economic Co-Operation and Development, *Greener and Smarter: ICTs, the Environment and Climate Change* (Paris: Organisation for Economic Co-Operation and Development, 2010), 19.

26. International Telecommunication Union, *Measuring the Information Society: Executive Summary* (Geneva: International Telecommunication Union, 2012), 4.

27. Maxwell and Miller, *Greening the Media.* Readers may wish to consult Basel Action Network and Silicon Valley Toxics Coalition, *Exporting Harm: The High-Tech Trashing of Asia* (Seattle: Basel Action Network, 2002); Sherry Lee, "Ghosts in the MACHINES," *South China Morning Post Magazine,* May 12, 2002, available at http://ban.org/library/ ghosts_in.html (accessed March 13, 2014); Xin Tong and Jici Wang, "Transnational Flows of E-Waste and Spatial Patterns of Recycling in China," *Eurasian Geography and Economics* 45, no. 8 (2004): 608–21; Zack Pelta-Heller, "HP's Printer Cartridges Are an E-Waste Disaster—Does the Company Really Care?," *AlterNet,* October 29, 2007, available at http://www.alternet.org/story/65945/hp%27s_printer_cartridges_are_an _e-waste_disaster_—_does_the_company_really_care (accessed March 13, 2014); Coby S. C. Wong, S. C. Wu, Nurdan S. Duzgoren-Aydin, Adnan Aydin, and Ming H. Wong, "Trace Metal Contamination of Sediments in an E-Waste Processing Village in China," *Environmental Pollution* 145, no. 2 (2007): 434–42.

28. Mike Hibberd, "Public Private Partnership," *Telecoms.com*, September 15, 2009, available at http://www.telecoms.com/14505/public-private-partnership (accessed March 13, 2014).

29. *ICTs for Environment: Guidelines for Developing Countries, with a Focus on Climate Change*, ICT Applications and Cybersecurity Division Policies and Strategies Department (Geneva: International Telecommunication Union, 2008); *ITU Symposium on ICTs and Climate Change Hosted by CTIC, Quito, Ecuador, 8–10 July 2009: ITU Background Report*, Telecommunication Development Sector (Geneva: International Telecommunication Union, 2009).

30. Hamadoun I. Touré, ITU Secretary-General's "Declaration on Cybersecurity and Climate Change, High-Level Segment of Council (Geneva: International Telecommunication Union, November 12–13, 2008), available at http://www.itu.int/council/C2008/hls/statements/closing/sg-declaration.html (accessed March 13, 2014).

31. John Houghton, "ICT and the Environment in Developing Countries: Opportunities and Developments," paper prepared for the Organization for Economic Cooperation and Development, 2009, available at http://www.oecd.org/ict/4d/44005687.pdf (accessed March 13, 2014).

32. Mark Mills, *The Cloud Begins with Coal: Big Data, Big Networks, Big Infrastructure, and Big Power: An Overview of the Electricity Used by the Global Digital Ecosystem*, National Mining Association/American Coalition for Clean Coal Electricity, 2013, available at http://www.americaspower.org/sites/default/files/Cloud_Begins_With_Coal_Exec_Sum.pdf (accessed March 13, 2014).

33. Greenpeace, *How Clean is Your Cloud?* 2012, available at http://www.greenpeace.org/international/en/publications/Campaign-reports/Climate-Reports/How-Clean-is-Your-Cloud (accessed March 13, 2014).

34. Duncan Clark, "Google Discloses Carbon Footprint for First Time," *Guardian*, September 8, 2011, available at http://www.guardian.co.uk/environment/2011/sep/08/google-carbon-footprint (accessed March 13, 2014).

35. George Ritzer and Nathan Jurgenson, "Production, Consumption, Prosumption: The Nature of Capitalism in the Age of the Digital 'Prosumer,'" *Journal of Consumer Culture* 10, no. 1 (2010): 13–36.

36. Richard Maxwell and Toby Miller, "'Warm and Stuffy': The Ecological Impact of Electronic Games," in *The Video Game Industry: Formation, Present State, and Future*, ed. Peter Zackariasson and Timothy Wilson (London: Routledge, 2012), 179–97; Toby Miller, *Cultural Citizenship: Cosmopolitanism, Consumerism, and Television in a Neoliberal Age* (Philadelphia: Temple University Press, 2007), and "Feminist Nudity? Video Consent? Moral Panic?" *Los Angeles Review of Books*, August. 23, 2012, available at http://lareviewofbooks.org/essay/feminist-nudity-video-consent-moral-panic (accessed March 13, 2014).

37. Cosma Orsi, "Knowledge-Based Society, Peer Production and the Common Good," *Capital & Class* 33 (2009): 31–51.

38. Erin Sparks and Mary Jo Watts, *Degrees for What Jobs? Raising Expectations for Universities and Colleges in a Global Economy* (Washington, D.C.: National Governors Association Center for Best Practices, 2011).

39. Nandini Lakshman, "Copyediting? Ship the Work Out to India," *Business Week*, July 8, 2008, available at http://www.businessweek.com/stories/2008-07-08/copyediting-ship-the-work-out-to-indiabusinessweek-business-news-stock-market-and-financial-advice (accessed September 20, 2014).

40. Dawn C. Chmielewski, "Poptent's Amateurs Sell Cheap Commercials to Big Brands," *Los Angeles Times*, May 8, 2012, available at http://articles.latimes.com/2012/may/08/business/la-fi-ct-poptent-20120508; and http://www.poptent.net (both accessed March 13, 2014).

41. Antonio Negri, *goodbye mister socialism* (Paris: Seuil, 2007); Alvin Toffler, *Previews and Premises* (New York: Morrow, 1983).

42. André Gorz, "Économie de la connaissance, exploitation des savoirs: Entretien réalisé par Yann Moulier Boutang and Carlo Vercellone," *Multitudes* 15 (2004), available at http://multitudes.samizdat.net/Economie-de-la-connaissance (accessed March 13, 2014).

43. Sharada Ramanathan, "The Creativity Mantra," *The Hindu*, October 29, 2006, available at http://www.hindu.com/mag/2006/10/29/stories/2006102900290700.htm (accessed March 13, 2014).

44. Andrew Ross, "Nice Work If You Can Get It: The Mercurial Career of Creative Industries Policy," *Work Organisation, Labour and Globalisation* 1, no. 1 (2006–07): 1–19; Carmen Marcus, *Future of Creative Industries: Implications for Research Policy* (Brussels: European Commission Foresight Working Documents Series, 2005).

45. Australian Academy of the Humanities, *Submission in Response to Research Workforce Strategy Consultation Paper: Meeting Australia's Research Workforce Needs*, 2010, available at http://www.innovation.gov.au/Research/Documents/Submission75.pdf (accessed March 13, 2014); Stuart Cunningham, "Oh, the Humanities! Australia's Innovation System out of Kilter," *Australian Universities Review* 49, no. 1–2 (2007): 28–30.

46. Stuart Cunningham, "Developments in Measuring the 'Creative' Workforce," *Cultural Trends* 20, no. 1 (2011): 25–40.

47. Hasan Bakhshi, Philippe Schneider, and Christopher Walker, *Arts and Humanities Research and Innovation* (London: Arts and Humanities Research Council / National Endowment for Science, Technology and the Arts, 2008).

48. See https://www.google.com.mx/search?q=london+series+monet&es_sm=119&tbm=isch&tbo=u&source=univ&sa=X&ei=4laKU7ugOYPYOp_ugMgJ&ved=0CDMQsAQ&biw=1306&bih=651; and https://www.google.com.mx/search?q=constable+clouds&es_sm=119&tbm=isch&tbo=u&source=univ&sa=X&ei=EleKU4CEKsOVPJHBgJgO&ved=0CCcQsAQ&biw=1306&bih=651 (both accessed March 13, 2014).

49. John E. Thornes, "A Rough Guide to Environmental Art," *Annual Review of Environment and Resources* 33 (2008): 391–411; Amy Struppek, "The Social Potential of Urban Screens," *Visual Communication* 5, no. 2 (2006): 173–88.

50. Charles Babbage, *On the Economy of Machinery and Manufactures*, 1832, available at http://www.gutenberg.org/ebooks/4238 (accessed March 13, 2014).

51. See http://thecreatorsproject.vice.com/blog/8-projects-turning-deadly-e-waste-into-beautiful-non-deadly-works-of-art (accessed March 13, 2014).

52. See http://www.jessicamillman.net/1/post/2010/10/e-waste-crisis.html (accessed March 13, 2014).

53. Priya Ganapati, "Old Hard Drives Get Sculpted into Cars, Bikes, Robots," *Wired*, December 10, 2009, available at http://www.wired.com/gadgetlab/2009/12/hard-drives-sculpture/3 (accessed March 13, 2014).

54. See http://www.sudhutewari.com/DumpArt/ST_DumpArt.html (accessed March 13, 2014).

55. See http://nomeedonna.deviantart.com (accessed March 13, 2014).

56. See http://www.chrisjordan.com/gallery/rtn/#unsinkable (accessed March 13, 2014).

57. See http://www.erikotto.com/exhibitions_03.htm (accessed March 13, 2014).

58. See http://ink-dwell.com/ (accessed March 13, 2014).

59. See http://www.whatech.com/members-news/green-technologies/18565-melbourne-hosts-australia-s-largest-ever-e-waste-artwork-bringing-to-life-our-throw-away-technology-addiction; and http://www.mobilemuster.com.au/news/articles/2014-02/blog-article-mobilemuster-presents-chris-jordans-australian-tour (both accessed March 13, 2014).

60. See http://www.yonafriedman.nl (accessed March 13, 2014).

61. See http://www.dirtstudio.com/; http://www.stacylevy.com (accessed March 13, 2014).

62. Natalie Behring, "Inside the Digital Dump," *Foreign Policy*, May/June, 2007, available at http://www.foreignpolicy.com/articles/2009/09/23/inside_the_digital_dump (accessed March 13, 2014).

63. See http://urbanscreens.org (accessed March 13, 2014).

64. See http://yurisuzuki.com/works/tube-map-radio (accessed March 13, 2014).

65. See http://petermcfarlane.com/Peter%20Macfarlane%20web%20site/03circuit_board.php (accessed March 13, 2014).

66. See http://ralonso.com/?portfolio=new-2&lang=en (accessed March 13, 2014).

67. See http://www.instructables.com/id/New-Robo-planters (accessed March 13, 2014).

68. See http://www.ewasteworkshop.com/category/ar/ (accessed March 13, 2014).

69. See http://www.daniploeger.org/#!wastecircuits/cvq4; http://www.daniploeger.org/#!electrode/c3kk (both accessed March 13, 2014).

70. See http://www.798district.com/ (accessed March 13, 2014).

71. See http://vimeo.com/10075678 (accessed March 13, 2014); Kavita Philip, "Art and Environmental Practice," *Capitalism Nature Socialism* 19, no. 2 (2008): 69–74, 70.

72. See http://www.nyu.edu/projects/xdesign/ (accessed March 13, 2014).

73. See http://www.francisalys.com/public/collector.html (accessed March 13, 2014).

74. See http://steinhardt.nyu.edu/faculty_bios/view/Natalie_Jeremijenko (accessed March 13, 2014).

75. Adrian Wooldridge, "The Coming Tech-Lash," *Economist*, November 18, 2013. See http://www.economist.com/news/21588893-tech-elite-will-join-bankers-and -oilmen-public-demonology-predicts-adrian-wooldridge-coming (accessed March 13, 2014).

76. See https://www.facebook.com/OccupySiliconValley408 (accessed March 13, 2014).

77. See http://anonnews.org/ (accessed March 13, 2014).

78. Georg Simmel, "The Philosophy of Landscape," trans. Josef Bleicher, *Theory, Culture and Society* 24, no. 7–8 (2007): 20–29.

79. Néstor García Canclini, *Art Beyond Itself*, trans. David Frye (Durham: Duke University Press, 2014).

80. Zulma-Lin Garcia-Morales, "E-Waste: A Growing Problem with Global Expectations," *Museums & Social Issues* 6, no. 2 (2011): 196–203.

81. William Morris, *Art and the Beauty of the Earth*, 1884; see http://www.marxists .org/archive/morris/works/1881/earth.htm (accessed March 13, 2014).

CHAPTER 7

Cellular Borders

Dis/Connecting Phone Calls in Israel-Palestine

HELGA TAWIL-SOURI

Telephonic (Im)Possibilities

Contradictory conceptions of borders, frontiers, buffer zones, and divisions have particular salience in the landscape of Israel/Palestine.[1] Depending on one's position and political status, a settler outpost, a Palestinian city, or a checkpoint can be easy for some to pass through or impossible for others.[2] These different territorial and political spaces (illegal towns, open-air prisons, strict border crossings nowhere near a border) mean that flows in, out, and around them are politically constructed to be uneven, depending on one's position.

   This chapter deals with two issues. First it takes something as benign as a telephone call—its underlying infrastructure, its political geography, and its political economy—and demonstrates how infrastructure is a dynamic manifestation of the tensions between Israeli practices of control and bordering on the one hand, and Palestinian attempts to mitigate or negate these on the other. In Israel/Palestine, the telecommunications infrastructure is not a metaphor for the conflict, it is the conflict in material form. Who can call what number on what network from what location and at what price are deeply political concerns shaped by the uneven relationship Palestinians and Israelis have to the construction and enforcement of territorial borders. The objective here is not simply to highlight the ways telecommunications is restricted but how they also follow territorial limitations and thus literally "mark" and "make" the territory as a result of historical-political-spatial processes.

Second, this chapter addresses the politicization of technology and the formations and negotiations of new kinds of borders. It shows how media infrastructures and networks—such as telecommunications—are not in and of themselves boundless and open but function as politically defined territorial spaces of control and are integral aspects of states' *territoriality*, bringing into stark question assumptions about globalization, communication, sovereignty, and borders—in Israel/Palestine and beyond.

In what follows, I focus on three specific locations and analyze the (im)possibilities of telephone calls between them. I begin with Migron, an Israeli settler outpost in the West Bank and eight kilometers inside the "Green Line."[3] Next, I move six kilometers directly east as the crow flies, to Ramallah, which has become the de facto capital, of, if nothing else, the West Bank, the Palestinian Authority, most international aid organizations in the Palestinian Territories, and which is increasingly billed as the successful city of neoliberal policies.[4] About five kilometers south of Ramallah (and ten kilometers north of Jerusalem) lies the Qalandia checkpoint, where I end. Qalandia emerged in 2000 and has since become a "terminal," in the language of the Israeli military, that serves to separate the southern parts of the West Bank (where the cities of Bethlehem and Hebron are) from the central West Bank, as well as the *entirety* of the West Bank from Jerusalem and its surroundings.[5] Rather than depicting the separation and inequalities between these locations through an analysis of land ownership, movement of people, or physical markers such as walls and highways, this chapter analyzes the telecommunications infrastructure that disconnects them.[6]

MIGRON

- Territorial/political status: illegal Jewish-Israeli settler outpost.
- Landline area code: 02, serviced by Israeli providers.
- Cellular access and area code, serviced by Israeli providers: Cellcom 052/053. Orange 054. Pelephone 050/051/056. MIRS 057. No Palestinian providers or signals.

Israel has one of the world's highest cellular penetration rates, 132 percent in 2011, and boasts a tied-for-sixth-place position with Denmark, Finland, and Norway for highest smartphone penetration.[7] By 2001 Israel was one of only two countries (with Luxembourg) to have passed the 100 percent cellular penetration rate threshold.[8] To say that cellular phone use and service across Israel are ubiquitous is to state the obvious.

Figure 7.1. Migron and its cellular towers. Photo by author.

In 1986 the national telecommunications company Bezeq launched its cellular subsidiary, Pelephone, which offered mobile service inside Israel and to Israeli settlements in the West Bank and the Gaza Strip. Since market liberalization in 1994 another three private cellular companies provide service: Cellcom, which launched in 1994, MIRS in 1998, and Orange in 1999.[9] The four providers contend that they operate in the Palestinian Territories so as to provide service to settlers, Israelis traveling on bypass roads, and, of course, the military. They generally have the liberty to install their equipment wherever they want inside the Territories, although the majority have been established inside settlements, military zones, and along bypass roads.[10] This means that Israeli infrastructure more or less parallels Israel's territorial presence in the West Bank (since disengagement from Gaza in 2005, Israeli firms no longer have equipment inside the strip).

Settler presence in the Territories vis-à-vis telecommunications poses a chicken-and-egg dilemma, however, in that there have been times when the telecommunications infrastructure was built *before* presence of settlers. For example, in the fall of 2000 Pelephone illegally installed a transmission tower atop a hill six kilometers directly east and in line-of-sight of the West Bank city of Ramallah. The company then pressured the Israeli government to install electricity lines in order to power the tower. It did not take long for a group of

Jewish-Israeli settlers to hook up five caravans to the electricity network and make the hill their home, disregarding the fact that the land belonged, even according to Israeli law, to Palestinian families from villages a few hundred meters away. A few months later, the Israeli Ministry of Construction paved a dirt road and installed streetlights. By March 2001 the Israeli military came to guard the illegal settlers. Ten years later, in 2011, the Migron outpost—whose electrified fence is lined with guard dogs, surveillance cameras, and a gate manned by the military 24/7—was home to more than fifty families. Inside the fence stood a Pelephone and an Orange tower beaming cellular service to Migron's residents, to nearby settlements, and to those driving along Highway 60 (a bypass road that is off-limits to Palestinians).

The story of Migron is significant. Something as seemingly benign as a cellular tower serves as the roots for territorial colonization. But the towers also mark a particular kind of land grab of a digital colonization process that combines territorial and high-tech presence and control, highlighting the paradoxes of (uneven) borders in the landscape of telecommunications infrastructures in Israel/Palestine. Neither Migron nor the presence of Pelephone and Orange is unique. *All four* Israeli providers have dozens of antennas, transmission stations, and additional infrastructure across occupied territory: as of 2009 MIRS owns about ninety antennas and communication facilities built on occupied territory, Cellcom at least 191, Pelephone 195, and Orange 165.[11] Migron was "evacuated" by the Israeli government in late 2012, moving the settlers a few kilometers down the street. The cellular towers inside Migron remain and are still beaming strong signals all around.

Although the Green Line was considered the de jure border in 1967, according to UN Resolution 242, Israel has long since trespassed it in (mis)appropriating Palestinian land well beyond it. Some scholars posit that settlements were the edifices that initially ruptured the 1967 boundary,[12] and that since then, and despite the 1991–1993 Oslo Accords, that rupture has expanded into a wider-reaching network that includes the territorial expansion of settlements and burgeoning settler population, the popping-up of outposts such as Migron, the shifting and growing matrix of bypass roads and checkpoints, military zones and "green areas" deep in the West Bank, the widening buffer zone inside the Gaza Strip, the enlarging of Jerusalem's boundaries, and, as in the example above, telecommunications infrastructure. While there exists a matrix that has seeped into Palestinian territory, the presence of Israeli cellular infrastructure, flows, and signals demonstrates the extent to which the boundaries of the Israeli regime are much more fuzzy, wide-reaching, and dynamic than traditionally understood territorial presence. Cellular signals by their nature do not "know" to stop at political boundaries. Given the location of settlements, outposts, and military areas—ubiquitously scattered and usually atop the highest points throughout the West Bank—Israeli operators' signals blanket much of the Palestinian Territories. In the case of Migron, Israeli cellular signals can be enjoyed throughout the nearby Palestinian villages and in the de facto capital of Ramallah, easily spotted when standing in Migron (see figure 7.2). As such, the breadth of Israeli presence and control over Palestinians exists throughout the territory of Israel/Palestine, seeping through multiple spaces of Palestinian life.

The four Israeli cellular providers are awarded spectrum from the Israeli Ministry of Communication. None of the providers has faced any difficulty in obtaining licenses or spectrum from the MoC. Pelephone has 46MhZ of spectrum, Cellcom 27MhZ, and Orange 20.4MhZ (MIRS, being the military's official and exclusive operator, does not make its spectrum allocation publically known).[13] The four companies provide all of the latest technologies to their subscribers: 4G, GPS tracking, online banking, and so on. Each provider is awarded its own area code(s), and providers must establish "bilateral agreements" (in corporate-speak) to connect different area codes and allow users to roam on

Figure 7.2. View of Ramallah from Migron. Two other settlements are on the two hills on both sides of Ramallah; a Palestinian town is in the foreground. Photo by author.

different networks. As such, a Migron resident who may be a Cellcom subscriber and whose cellular phone number begins with the 052 area code can seamlessly call a cellular number of a friend in Tel Aviv on the Orange network (area code 054), a landline number in Jerusalem (area code 02), a landline number in the Ofra settlement up the street (area code 02), or a MIRS cellular number (area code 057) whose user happens to be driving along a bypass road somewhere in the Northern West Bank. In short, our Migron resident is unencumbered by area codes, cellular or landline numbers, or the geographic presence of a mobile user. He may be charged differently by Cellcom whether he is calling a landline or cellular number, and he may be charged more for any roaming charges should his Cellcom signal be weak and automatically switched to another provider's network, but he has the ability to make all such calls seamlessly and at local rates (in other words, there is no extra charge for calling across the Green Line). Moreover, should he wish to establish a landline inside his caravan, he would be able to do so through one of the six Israeli landline providers and obtain a 02 area code. The difference, telephonically, between our Migron resident and his Jewish-Israeli counterpart in West Jerusalem or Tel Aviv is nil.

For the Palestinian resident living within earshot of Migron, making *any* kind of telephone call—landline or cellular—is much more complicated, if possible at all. The Israeli-Palestinian technological relationship, like their political and economic relationship, has been one of Israeli control and restrictions and Palestinian dependence. From the outset of occupation in 1967, Israel controlled and maintained telecommunications systems in the occupied Territories and imposed legal and military restrictions. In terms of landlines, despite the fact that Palestinians paid income, value-added, and other taxes to the Israeli government, Bezeq was neither quick nor efficient in servicing Palestinian users. By the early 1990s only 2 percent of Palestinian households in the West Bank and the Gaza Strip had functioning landlines, and about 10 percent were connected to the network. Suffice it to say that, telephonically, Palestinians were enclavized and largely disconnected from the infrastructure, living under a regime that restricted both their mobility and their access to the outside world. Cellular telephones were not permitted at all, under a military rule imposed in 1989 that had also prohibited the use of telephone lines for the sending of faxes, emails, or "any form of electronic posting" from the Territories.[14] Nor were Palestinians permitted to build or have their own infrastructure. In fact, what little had been done with regard to telecommunications in Palestinian areas rendered the network subservient to infrastructure within Israel-proper (in other words, on the other side of the Green Line). All telephone-switching nodes were built *outside* areas that might eventually have to be handed over to Palestinian control; thus, calls between Ramallah and Nablus, for example, were connected in Afula.

Circumventing regulations on landlines was impossible—if the town was not connected to the network, there was simply nothing to do about it; if the town was connected but Bezeq did not connect the household, or took ten years to do so, nothing could be done about that either. Cellular telephony provided more loopholes. The MoC and military government had stipulated that Palestinians were not allowed to have cellular phones, but not that Israeli providers could not service the Territories. The difference was on the level of the *individual*, not the *territory*. Israeli signals—and by extension technological and economic flows—would be permitted practically anywhere in the territory of Israel/Palestine. Furthermore, the ban on cellular use was restricted to subscribers; thus nothing prevented Palestinians from purchasing Israeli telephones and buying pay-per-use cards. The Israeli cellular providers did not do anything to stop this: they were making substantial profit from Palestinian cellular use. From the Palestinian user's perspective, Israeli cellular reach was effectively boundless with Israeli signals available throughout the occupied Territories, and no authority was preventing pay-per-use service. In short, it made sense to have Israeli cell phones—never mind that there was simply no alternative. Until 1999—when a Palestinian cellular provider was first established, detailed below—there were approximately one hundred thousand Palestinian cellular customers of Israeli operators in the West Bank and the Gaza Strip. Palestinian territory was neither bounded nor bordered for Israeli providers. Such "boundlessness" would become politically and economically critical in the aftermath of the peace accords.

RAMALLAH

- Territorial/political status: Occupied city. De facto Palestinian capital; Area A.
- Landline area code: 02, serviced by Paltel (through Bezeq or other Israeli providers).
- Cellular access and area codes: Palestinian carriers only, who must obtain permission from Israeli Ministry of Communications: Jawwal/059. Wataniya/056. Israeli carriers accessible illegally.

The agreements of Oslo I (signed in 1993) and Oslo II (signed in 1995) would reverse many of the restrictions imposed on Palestinians. Palestinians were promised direct domestic and international telephone and Internet access and given permission to establish their own infrastructures. Oslo II specified: "Israel recognizes that the Palestinian side has the right to build and operate separate and independent communication systems and infrastructures including telecommunication networks."[15] It then went on to stipulate, however, the conditions within which an "independent" Palestinian infrastructure would be constrained:

> The Palestinian side shall be permitted to import and use any and all kinds of telephones, fax machines, answering machines, modems and data terminals. . . . Israel recognizes and understands that for the purpose of building a separate network, the Palestinian side has the right to adopt its own standards and to import equipment which meets these standards. . . . The equipment will be used *only* when the *independent* Palestinian network is operational.[16]

That the network would become independent *only* when the system became operational is crucial, because the Palestinian network to this day is not independently operational and continues to rely on Israel's in a catch-22 logic. As with other infrastructures (for example, broadcasting, sewage, population registries, water, transportation), Palestinians were subject to Israeli constraints that would counter their right—or simply their ability—to build separate and independent systems. With regard to telecommunications, Israel continues to determine the allocation of frequencies, where Palestinians are permitted to build infrastructure and install equipment, and much else that shapes the field.

Israel handed over responsibility for telecommunications in 1995 to the Palestinian Authority (PA). What little there existed of a technically debilitated fixed-line infrastructure in permissible areas was handed over; in the remainder of Palestinian territory, the PA would be responsible for building it from the ground up. The PA began to establish a simulacrum of an "independent" telecommunications system and awarded the newly formed Palestinian telecommunications company, Paltel, an exclusive ten-year license to operate fixed-line systems and a twenty-year contract to run mobile services. The license permitted Paltel to build, operate, and own landlines, a GSM (global system mobile communications) cellular network, data communications, paging services, and public phones. While Palestinian telecommunications infrastructure building, development, control, and use were now permitted, it would neither exist nor develop without continued Israeli-imposed limitations. Just as the geographies of the West Bank and the Gaza Strip were increasingly fragmented and contained during the post-Oslo "peace years" by Israeli expansion of settlements, settlers, checkpoints, walls, bypass roads, and the like, the allowable space of communication infrastructure was also confined to follow territorial boundaries. But only for Palestinians. For as the case of Migron demonstrates, Israeli infrastructure, networks, and signals are unfettered relative to Palestinian ones.

Palestinian infrastructure was—and continues to be—permitted to be built, accessed, and maintained only in the Oslo-defined Areas A and B (10 percent and 40 percent of the West Bank, respectively).[17] What this translated into was a fragmented network that had to physically circumvent more than 60 percent of the West Bank and 40 percent of the Gaza Strip.[18] Thus, landline telephone networks, cellular networks, and Internet connections invoke the parcelization and fragmentation of the Territories themselves.

Telecommunications, and Paltel especially, were celebrated as signs of successful state-building and hailed as the proto-state's entrance into the global network age. Indeed, Paltel was one of the first functioning national institutions. There would, however, be multiple ways in which Paltel—and telecommunications generally—would not be "national." First, telecommunications did not belong to the people of the nation; it was a private, for-profit enterprise. Second, it would continue to be reliant on Israeli infrastructure. All of Paltel's international calls, whether incoming or outgoing, would continue to be routed through Israeli providers because Paltel was not permitted its own international gateway. All of Paltel's Gaza–West Bank calls would be switched inside Israel because Paltel could neither dig under Israeli land to install a fiber-optic cable nor be allocated enough spectrum bandwidth to use microwave technologies. Paltel calls within the West Bank and within the Gaza Strip would also frequently be routed through Israeli providers because of the limitations of where and what kinds of equipment Paltel could install. Third, therefore, telecommunications would also never be territorially national, in that not all parts of the Palestinian Territories would be wired. Fourth, in the realm of policy decisions, the Palestinian Ministry of Telecommunications (MPTT, later transformed into the MTIT) would be constrained by the Israeli Ministry of Communication and the rest of the Israeli regime's occupation apparatus. Finally, the PA's policies would not challenge Israel's ultimate control over and containment of infrastructure. The reliance on Bezeq for much of their national connections and for all international connections would not end with the advent of Paltel. As Bezeq spokesman Roni Mandelbaum quipped in 1996, Palestinians "are not entitled to any signs of sovereignty. . . . They have to rely on the infrastructure we supply them."[19] This has yet to fundamentally change. The only "sovereignty" gained by Paltel was due to the liberalization of the *Israeli* market when Paltel could choose between different Israeli providers. Like much else in the post-Oslo era, Paltel was a "national institution" within the confines of Israeli control.

Unlike landline infrastructure, cellular telephony in Israel and Israeli controlled Territories was largely driven by commercial growth and existed primarily in Areas C and Israeli-controlled areas such as settlements and outposts; thus, none of the existing cellular network in any Palestinian territory was handed over in 1995. Paltel had to establish its own cellular infrastructure within the confines of Area A and parts of Area B. Paltel's cellular subsidiary, officially launched in 1999 under the name Jawwal, would also be bound. Everything would be determined by Israeli permissions, from the strength of transmission towers to the kinds of routers and switches necessary to enable cellular traffic, from spectrum allocation to the location of equipment; some of these limitations would simply be imposed by military officials or by the MoC; others, as detailed in annexes of the Oslo Accords, would have to be agreed upon in a bilateral body, the Joint Technical Committee (JTC), in which Israel would have power to veto. For example, with respect to GSM and other cellular frequencies, "mutual participation will be agreed in the JTC according to the planning of each side, and the division of . . . [cellular] frequencies will take into account the users ratio of each side."[20] Given the limitations imposed on the Palestinian infrastructure to begin with, the Palestinian user ratio would be forced to remain lower, thus continuously justifying why Palestinians were awarded less spectrum frequency. Jawwal was awarded 4.8MhZ of spectrum at the 900 MHz range—less than any other cellular provider in the world (by comparison, the largest Israeli cellular operator, Pelephone, enjoyed 46MhZ of spectrum allocation).[21] The annex stated that "frequencies will be assigned upon specific requests" or be assigned "as soon as any need arises"—in both cases, to be decided by the Israeli side.[22] This seemingly simple but extremely important point became a limiting and bordering mechanism faced by Palestinian cellular operators and users for the years ahead.

Telecommunications frequencies and spectrum allocation would remain contingent on "final status issues" in ongoing and often frozen negotiations. The range of the electromagnetic spectrum was also determined by the MoC, so that "any future expansion was difficult to achieve [for Jawwal] and raised the costs of network equipment needed,"[23] as Jawwal would have to install more towers (at lower heights and with weaker signals) to cover a particular area and ensure service for a certain number of users. Constraining issues that limited the building or growth of fixed lines existed in the cellular realm, too: the forbidding of international access, the determining of regional codes, having to submit requests to the JTC for most needs and demands, and so on. The building of any cellular infrastructure in Area C and parts of Area B could only be achieved with Israeli permission, which "in most cases . . . are denied."[24] In the words of Jawwal CEO Hakam Kanafani, "The unique political situation [in Palestine] . . .

means that, unlike any other provider, Jawwal's network expansion is not only linked to financial and demand attributes, but also to the decisions of a foreign government."[25] The limitations often translated into higher setup costs for Jawwal: having to build more towers, not being able to pass directly from point A to point B, having to set up two separate operating entities in the West Bank and Gaza in terms of equipment and employees. Jawwal's first-phase capacity was 120,000 subscribers—a number limited by the combination of technical aspects of switching equipment, the number and strength of transmission towers, and spectrum allocation. The combination of a smaller sliver of frequency spectrum, limitations on equipment, and limitations on where and how strong antennas and base stations could be effectively limited both the number of subscribers Jawwal could simultaneously serve and where cellular users could obtain a signal.

By the time Jawwal began operation in 1999, more than one hundred thousand Palestinian users were already on Israeli networks. As the four Israeli providers had already established presence throughout the Territories, operated in a competitive landscape, and enjoyed 2,000 percent more frequency spectrum than Jawwal, it was all the more difficult for Palestinians to give up using Israeli services, even if increasingly labeled as antinationalistic.

The legal landscape had changed, but the practices continued. The Oslo Accords stipulated that each side's providers would not interfere with the other: "Both sides shall refrain from any action that interferes with the communication and broadcasting systems and infrastructures of the other side."[26] According to the terms set forth by Oslo II, the PA-designated provider (in this case Paltel and its subsidiary Jawwal) was to be the only cellular provider in the Territories, while Israeli providers would continue to serve the settlements.[27] It was illegal, according to MoC regulations and to the newly established Palestinian telecommunications law, for any provider, Israeli or otherwise, to operate in the Territories without the legal protocols for doing so: obtaining a license from the PA and paying taxes to the PA (and obtaining spectrum allocation and an area code from Israel's MoC). But, for the four Israeli cellular companies, such territorial, legal, and political constraints remained largely insignificant. The Israeli providers had unlicensed distribution and sales points in the Territories, did not operate with permits from the PA, nor obliged the Palestinian economy—whether in the form of license fees, taxes, or hiring of Palestinian employees. In short, their operation inside the Territories became illegal (and Palestinian users on those networks continued to be illegal, now also according to Palestinian law). In the words of Jawwal's first CEO, Hakam Kanafani:

> Jawwal's starting point was unlike that of any GSM provider in the region and possibly, in the world. . . . In most countries, the first GSM operator is granted

exclusivity for a number of years, during which the operator is expected to introduce its services, reach the breakeven point, and at the same time educate and prepare the market for a second operator. . . . This scenario did not work in Jawwal's case.[28]

Another Jawwal executive summed it up simply: "The Palestinian cellular market is a cost-free market [for Israeli providers]."[29] Israeli providers did not build, install, or maintain any *more* infrastructure than what they already needed to serve settlements and outposts, yet they benefited financially from Palestinian customers/use. Although their presence was in violation of Israel's telecommunications policies, of the PA's, and of Oslo's, they benefited—and continue to benefit—from Palestinian use and never tried to thwart it. This is an economic issue that stems from the uneven relationship between Palestinians and Israel: Palestinians are financially bound to Israel, as a "captive market" for Israeli firms and as an economic "dumping ground" for Israeli goods. But it is also a symbolic, territorial, and political strategy of bounding Palestinians while simultaneously minimizing territorial borders or limitations on Israeli flows—whether financial or technical.

The cellular system is constrained (or bordered) by the inherent design and limitations of the technology itself, as well as its relationship to other technical aspects such as frequency allocation, bandwidth, and transmission power. There is a territorial determination to these technical borders in how far signals can travel. Here, the borders that are enforced on Palestinian cellular flows are multifold, some inherent in the technical system itself (a signal can reach only so far), others imposed by legal and political decisions on the part of Israel. All of these result in a bounded cellular space for Palestinians.

Around the time Migron came into being, a person in Ramallah could legally purchase a Jawwal cellular phone and obtain the 059 area code. The euphoria of being able to support one's own national company made up for the worse-than-Israeli signals and higher-than-Israeli prices. Jawwal's sales had increased to more than one hundred thousand subscribers, in many cases driven by the violence, curfews, and closures of the Second Intifada, which began at the end of September 2000. A Jawwal user could call other Jawwal users, as well as anyone with a landline within the Palestinian Territories (which by this time had risen to about 9 percent of households, thanks to Paltel, compared with 43 percent in Israel). Making a call to a landline in Ramallah required one to dial the 02 landline code. Migron was also under the 02 area code; but a Jawwal user could not call those numbers—perhaps this is sensible, since interactions between Palestinians in the West Bank and settlers in Migron are nonexistent. But the inability to make that call was a policy/political decision on the part of Israel's MoC. A Jawwal number could not connect to any Israeli cellular number either. Any friend who had a Pelephone, for example, could not be reached from Jawwal. The firms did not have "bilateral agreements." In fact, what became rather common was for Palestinians to have two cellular phones: one Jawwal and one on an Israeli network.

Jawwal's coverage was understandably limited when it first launched, particularly in the West Bank. The fact that the West Bank's topography is hilly certainly did not help—Jawwal would have to install more towers in more places in order to reach valleys and hilltops and thereby further increase its operating costs. Jawwal was also constrained by the strength of signals. In most of the areas outside of downtown Ramallah, for example, Jawwal users simply had no signal. Over time, subscribers found it surprising that their lack of "bars" didn't increase. What those subscribers ought to have realized was that Jawwal had to work around technical limitations imposed not only by spectrum frequency and signal strength but also by the location of settlements, outposts, and Israeli cellular towers. From many parts of Ramallah one simply had to look up to the surrounding hills to understand why Jawwal's signals never arrived (see figure 7.3).

Figure 7.3. Ramallah with Psagot settlement and its broadcasting and cellular towers in the background. Photo by author.

At the end of 2000 Jawwal signed an agreement with Orange to share each other's transmission network. The deal made it possible to call an Orange number from a Jawwal number, and vice versa. Dan Eldar, vice president at Partner (Orange's parent firm), exclaimed, tongue-in-cheek, with reference to political negotiations, "I think we can say it's a bilateral agreement."[30] There wasn't much bilateral about the agreement, since Orange didn't need Jawwal's signals in the Territories. For Jawwal the primary reason for these agreements was to provide service in *Palestinian* areas where Jawwal is forbidden to build its network. Jawwal still does not have such agreements with Pelephone, Cellcom, or MIRS.

More and more people subscribed to Jawwal, largely driven by nationalist—not economic or technical—logic. But as Jawwal's network had to handle more subscribers, even though it continued to install new equipment wherever it was permitted, it would continue to be bound by the conditions under which it had first emerged. By 2007 it had 825,000 subscribers; 1.5 million by 2009; and more than 2 million by 2011—but its network *still only supports* its original 120,000.[31] Spectrum allocation increase has yet to be approved for Jawwal, restrictions on equipment continue to be draconian, and location of both Jawwal and Israeli infrastructure continues to be determined by the logic of Israeli occupation. Moreover, Palestinian users are still not permitted to have 3G (let alone 4G), GPS, online banking, and many other new mobile technologies and services because of Israeli policies against them. In December 2009, after four years of delay, Israel's MoC granted permission and spectrum to a second Palestinian provider, Wataniya. Both Wataniya and Jawwal operate under the same conditions. While the presence of two (legal) providers has helped drive prices down, the use of Israeli cellular phones has not decreased, for obvious reasons. In 2012 more than 2.5 million subscribers were on the Jawwal or Wataniya network, with a huge majority of them (more than 2.2 million on Jawwal) and approximately another 1 million Palestinians inside the Territories on Israeli networks. Market share for Jawwal and Wataniya combined is approximated to be between 60 percent and 80 percent of total Palestinian cellular use, as many Palestinians continue to rely on Israeli providers, either solely or in combination. The numbers are impossible to calculate with certainty because Israeli providers do not share that information and claim that they cannot know who is a pay-per-use subscriber (see figure 7.4 for a comparison of total number of subscribers on Israeli and Palestinian networks).[32]

The geography and the *control over* the geography of the Territories makes it possible for Israeli providers to service many parts of Area A and B: they are permitted to install antennas and base stations, and their antennas and cells in Area C have a wider range (thanks to being at higher elevations and stronger signal powers). Moreover, because Israeli providers enjoy a wider spectrum

Figure 7.4. Cellular subscribers on Israeli and Palestinian networks. Graphic by author.

allocation (see figure 7.5), they can handle more subscribers and simultaneous calls per cell. In other words, Israeli signals do not stop at the territorial boundaries imposed on Palestinians, but rupture them, reaching wider ranges before signals fade or are lost. A contradiction emerges about technical borders and to what extent they ought to follow or trespass territorial borders, and for whom.

Israeli cellular signals are exempt from boundaries and exempt themselves—illegally—from any responsibilities, financial or otherwise, toward the PA and Palestinians in general. Israeli cellular telephony functions not only according to the logic of "economies of scale" (the cost advantages a business obtains due to expansion) but economies of spectrum and economies of *spread* or, perhaps more appropriately, economies of digital colonization. Given that Jawwal's network expansion is beset by various kinds of limitations, its own network coverage lags well behind that of Israeli providers. Digital borders are erected for Palestinian providers and cellular users but not for Israelis. Israeli cellular space is guarded by bordering and bounding any Palestinian presence of "rupturing" or trespassing into it—Jawwal is not permitted to operate in Migron or

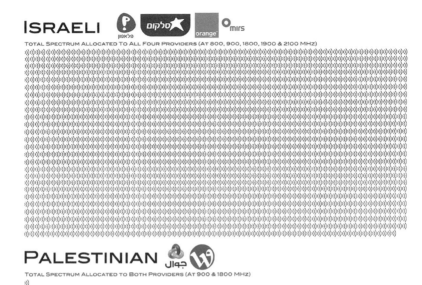

Figure 7.5. Spectrum allocation comparison for Israeli and Palestinian cellular providers. Graphic by author.

Tel Aviv, for example. Thus, borders on and for Palestinians are meant to trace territorial borders imposed by Israel that keep Palestinians contained. Jawwal's containment is determined by real territorial borders. Jawwal (and Paltel, as described earlier, as well as Wataniya) cannot erect antennas, transmission towers, or other equipment in more than 60 percent of the West Bank (*and* all of Israel). Cellular flows are determined and bound by the territorial landscape designed in the Oslo Accords: fragmented, enclavized, contained.

QALANDIA

- Territorial/political status: Israeli-manned military checkpoint. Area C.
- Landline area code: 02. Serviced by Bezeq until 1995; under Palestinian "control" since then, but no new landlines have been allowed since.
- Cellular access and area code: legally Palestinian, but no access available. Israeli carriers accessible illegally: Cellcom 052/053. Orange 054. Pelephone 050/051/056. MIRS 057.

The Oslo Accords territorially defined *where* telecommunications infrastructure could be built and set the context through which Israel could limit the kind of infrastructural equipment used. While service continues to be substandard or unavailable in many parts of Areas A and B, as described above, when it comes to the remaining 60 percent of the West Bank (Area C), it is more dismal. The accords had specified that "in Area C, although powers and responsibilities are transferred to the Palestinian side, any digging or building regarding telecommunications and any installation of telecommunications equipment, will be subject to prior confirmation of the Israeli side."[33] In other words, the PA would be given responsibility for servicing Area C, but not necessarily permission. This was reflective of the pseudosovereignty at the core of the Oslo Accords, of the contradiction of absolving Israel of "responsibility" yet keeping it as controlling power. If Area C were not serviced—whether for telecommunications, postal services, electricity, sewage, or otherwise—this would no longer be Israel's problem. Although the Oslo Accords have been officially "over" since 1998—they were five-year interim agreements—Palestinian telecommunications infrastructure continues to be territorially determined by Oslo's maps *and* by the Israeli regime's territorial changes in the form of settlement growth, bypass roads, buffer zones, checkpoints, walls, and the concomitant shrinking of Palestinian spaces.

Ironically, if there is any one place where Palestinians might actually need a cellular phone, it is at places like checkpoints, given that these dot the landscape, have a completely illogical and obscure system of letting people through or not, and are sites of physical violence on the part of Israeli soldiers toward Palestinians. One such checkpoint, only a few kilometers from Ramallah, is Qalandia. Qalandia is more than simply a checkpoint, though. It has for all intents and purposes become a border terminal that separates various parts of the West Bank from each other, as well as the entirety of the central and northern parts of the West Bank from Jerusalem and Israel. Qalandia has also become a central Palestinian transportation hub: it is the place through which all buses and long-distance taxis leave and arrive from all parts of the West Bank (to go from Hebron to Nablus, for example, one must switch buses at Qalandia). At its busiest periods, the checkpoint is a "station" for more than twenty thousand Palestinians per day. Some remain stuck for hours, some get arrested, some are physically hurt, some turn back.

Being in Area C, Qalandia falls under the responsibility of the PA but the strict "security"/military controls of Israel. Jawwal and Wataniya have yet to be given permission to install any equipment there or have their nearby towers beam strong-enough signals to provide service in and around the checkpoint. A Palestinian cellular user cannot call anyone from Qalandia—neither another cellular user on the same network nor one on an Israeli network with whom the Palestinian providers have roaming agreements. That Qalandia is in Area C also means that Paltel faces limitations on installing landlines, and for the majority of the households and businesses around the checkpoint, there is no fixed-line service either. A call between Qalandia and Ramallah, five kilometers apart, is impossible.

Other locations in Area C—for example, the villages around Migron—enjoy Israeli signals. No such signals are available at Qalandia. Israeli signals are not available there, either, for the simple reason that Israelis do not travel through the area.[34] Qalandia is a telephonic no-man's-land—quite appropriate, since it is, from an Israeli perspective, also a political and territorial no-man's-land, despite being a busy and bustling location.

* * *

The combination of Israeli policies territorially and "ethereally" constrain Palestinian cellular communications. On the ground there is no sovereign Palestinian communications infrastructure, and what exists is fragmented, dispersed, and often disconnected, as well as technologically stunted and overburdened. In many places it is simply nonexistent. Palestinian users and the infrastructure as a whole are territorially (and otherwise) bound by area codes, the landline infrastructure, the kinds of equipment permitted, and the range, strength, and direction of signals, among other policies, all of which follow the narrow and fragmented territorial boundaries of land enclosures.

Telecommunications infrastructures demonstrate the ongoing importance of territoriality—for Palestinians, for Israel, and more generally. Territoriality, and concomitant aspects such as bordering mechanisms, flows, and (im)mobilities, are products of social and material practices, themselves marked by uneven (re-)developments. We have not at all reached the age of the "end" of borders or the decreasing importance of territoriality in a state's power. Rather, practices of bordering and the continued importance of territoriality mark and stratify territory, people, and flows through different mechanisms. Infrastructures and networks—whether telecommunications or otherwise—are not open, liberatory, de-territorial, and borderless spaces, or certainly not so by "nature," but represent designed technical activities that are outcomes of social, economic, political, and territorial processes. They can very much function and be made to function as spaces of control and containment.

Territorially defining communication flows is not simply a matter of ensuring control but of bounding, defining, limiting, surveilling, and controlling Palestinian (communication) flows, period. There is, of course, an integral aspect of revenue streams, as described above, due to the reliance on the Israeli backbone. More important here is the issue of Israel's drive to secure all of its different kinds of borders from Palestinian "ruptures" and trespasses and simultaneously to ensure the containment of Palestinians in the technological and communication realm, as well as the economic, political, territorial, and symbolic realm.

Qalandia is not on any telecommunications network. In a sense, it doesn't exist. It is the same on Israeli-made maps: Qalandia is nowhere to be found. One can argue that it is, after all, "merely" a checkpoint. But what then can we make of Jerusalem? Jawwal and Wataniya do not have permission to exist in Jerusalem, as the entirety of the city's municipality (itself extended on Israel's maps and its territorial practices) is considered Israeli territory by Israel, and neither of the two Palestinian cellular companies (nor Paltel, for that matter) are permitted to provide services in Israel. Palestinians in Jerusalem have to rely on Israeli cellular providers and fixed-line providers. Jerusalem is part of area code 02, like Migron and Ramallah. However, while a call from Migron to Jerusalem is an ordinary occurrence, just as a call between Tel Aviv and Petakh Tikva is, for example, all calls from Palestinian Territories to Jerusalem are considered *long-distance international* calls. Jawwal, Wataniya, and Paltel are billed surcharges for connections to Jerusalem, just as the rest of Israel, by the Israeli providers. In a way, calling Jerusalem is no different than calling England and having the provider be surcharged by BT. There are no alternative means of connecting; everything is dependent on the Israeli backbone. Jerusalem, then, is in a foreign country: an unattainable capital for Palestinians. This enforced disconnection goes even further. Although the Palestinians were provided their own international dialing code (970) by the ITU in June 1999, despite Israeli objections at all levels up to the prime minister, Paltel is not permitted to install its own international exchange router and continues to rely on the Israeli backbone for *all* incoming and outgoing telephone calls. Area Code 970 is *always* routed through 972 (Israel's international code). Israel enjoys four international switching nodes and has direct links to global undersea telegraph and telephone cables: it is part and parcel of the global network. From the global network's perspective, 970 is not simply cosmetic: it suggests altogether that Palestine does not exist.

Im/Mobility

The borders of the technological may be less visible than the walls, gates, fences, and checkpoints of the physical world, but they are no less real and significant politically. What the Palestinian/Israeli case showcases is how decisive borders continue to be, how their related processes are shifting and dynamic, and how they are enforced, experienced, and circumvented in different ways and across different spaces. The fragmentation and diffusion of borders lies in the realm of the technical order we create and disseminate; thus, the other side of today's territorial transformation bears witness to massive fragmentations of landscape and the production of hermetic spaces and territorial, legal, and technological islands. The point is not simply to argue for a flip side of the borderlessness of the network age but to recognize that territorial borders are not only *increasing* (as they are in Israel/Palestine) but are equally manifested in the realm of infrastructure—digital, high tech, media, and all other forms.

Furthermore, that Palestinian technology infrastructures are constrained by Israeli policies demonstrates the spatial reach of Israel's power—well beyond any supposed territorial boundaries. Israel enjoys a monopoly on where to draw and how to secure its shifting borders. It is not a matter of where we are to place the boundaries geographically—the Green Line, around Area A, along the wall and buffer zones, around settlements—but of recognizing that borders are drawn unevenly on Israeli versus Palestinian flows. This invisible yet ubiquitous border enables great elasticity in the territorial aspects of sovereignty. This is a challenge to our conceptions of what it means to live in a global digital new order. We continue to assume stable points of view, a world of places, boundaries, Territories rooted in time and bounded in space;[35] but these spaces have their own (sometimes new) grammar that produces infrastructural contexts that result in uneven immobilities.

We recognize that technology infrastructures are actively involved in the production of space.[36] But the territorial aspect is slightly more complicated and must be understood in relation to im/mobility. Scholarship that has focused on cellular and mobile phones continues to argue that mobile phones *free* us from much spatial fixity and give rise to what might be called "networked individualism." Yet, as the case of Israel/Palestine demonstrates, mobile phones are constrained by spatial fixities of the infrastructural materiality determined by Israeli interests, and, simultaneously as they intersect—in a segregated manner—with Israeli mobile flows which themselves follow a territorial logic. While Palestinian mobile phone users can carry their phones around with them (and thus can be considered "mobile"), how far signals reach and where the infrastructure of Palestinian cellular networks reach are territorially defined by the logic of occupation.

Telecommunications networks are integral instruments in the production of new spatialities. The case of Migron/Ramallah/Qalandia (and Jerusalem) demonstrates how cellular telephony brings into question the political, territorial, and economic fixity and containment of (im)mobility. Mobility, like power, is highly differentiated and relational.[37] What exist are dynamic and contingent mobilities. As such, the relative immobilities enforced on Palestinian telecommunications flows (taken together with mechanisms such as checkpoints, bypass roads, walls, settlements, and so on) must be seen in relation to the relative mobilities they create for Jewish-Israelis. Migron's signals exist because Ramallah's are constrained in very real and territorial ways. Qalandia is not part of the infrastructure because it is not a space that Jewish-Israelis pass through. Palestine does not exist on the network because 970 is in actuality 972. The issue, however, is not simply to juxtapose Jewish-Israeli/settler mobility with Palestinian immobility but to recognize that living with and through im/mobility is a crucial and historically longstanding issue for all Palestinians. Being Palestinian is having to live with, negotiate, challenge, and resist various mechanisms and power struggles over movement and sedentariness. Being Palestinian means having to negotiate an unevenly marked and made territory and spatiality that are trespassable for Jewish-Israelis and bound and constrained for Palestinians.[38] The specifics of Palestinian landline telephony are an example of the contemporary processes of territoriality, border making, and an example of the segregation of a network and the processes of a seemingly ethereal but also very territorial *im*mobility.

Notes

1. Scholarship addressing territorial and symbolic borders in Israel/Palestine includes: Ghazi Falah and David Newman, "The Spatial Manifestation of Threat: Israelis and Palestinians Seek a 'Good' Border," *Political Geography* 14, no. 8 (1995): 689–706; Eyal Weizman, *Hollow Land: Israel's Architecture of Occupation* (New York: Verso, 2007); Helga Tawil-Souri, "Uneven Borders, Coloured (Im)mobilities: ID Cards in Palestine/Israel," *Geopolitics* 17, no. 1 (2012).

2. By "position" I mean where one happens to be standing, but also, more important in this case, what access to what network a person may have that is in itself dependent on one's citizenship and/or ethnicity. In the case of citizenship, Arabs and Jews who are Israeli citizens are provided uneven access to circuits of civility by the Israeli state (such as education, housing rights, travel permits); in the case of noncitizens, Palestinians in the Palestinian Territories or in and around Jerusalem are barred altogether from circuits within the Israeli state.

3. The "green line," the 1967 borders, and the 1949 Armistice Line are synonymous, demarcating the boundary between Israel and the Territories it captured and occupied in the 1967 war. It is important to note here that there are no roads that connect Migron to Ramallah; as an outpost—and similar to all settlements in the West Bank—Migron is connected to "Israeli" sites—inside the West Bank and outside—through a network of "bypass" roads, roads open only to Jewish Israelis. Moreover, a resident in Migron would not ever have to pass through, or near, the Qalandia checkpoint either, but he or she can enjoy a direct link to Jerusalem open only to Jewish Israelis. The Ramallah-

Jerusalem road, which has existed for decades, has been severed by Qalandia since 2000; however, Palestinians in the West Bank do not generally obtain permission to travel to Jerusalem; as such, while Qalandia does separate the West Bank from Jerusalem, it is also the point of separation of different parts of the West Bank from each other.

4. See Nasser Abourahme, "The Bantustan Sublime: Reframing the Colonial in Ramallah," *City*, 13, no. 4 (2009): 499–509; Lisa Taraki, "Enclave Micropolis: The Paradoxical Case of Ramallah/al-Bireh," *Journal of Palestine Studies* 37, no. 4 (Summer 2008): 6–20.

5. For a background on the Qalandia checkpoint, see Helga Tawil-Souri, "New Palestinian Centers: An Ethnography of the 'Checkpoint Economy,'" *International Journal of Cultural Studies* 12, no. 3 (2009): 217–35; Helga Tawil-Souri, "Qalandia Checkpoint as Space and Non-Place," *Space and Culture* 14, no. 1 (2001): 4–26.

6. See Weizman, *Hollow Land*; Rafi Segal and Eyal Weizman, eds., *A Civilian Occupation: The Politics of Israeli Architecture* (New York: Verso, 2003).

7. ITU (2013), *ICT Fact and Figures*, and Bruce Sterling, "42 Countries Ranked by Smartphone Penetration Rates," *Wired*, December 16, 2011.

8. ITU, *Stat Shot* 7 (August 2011), available at http://www.itu.int/net/pressoffice/stats/2011/03/#.Unq8viRA9PA.

9. Cellcom is part of the IDB Group, a conglomerate of Israeli and international companies whose largest shareholders are Motorola and Israeli IT firm Tadiran. Orange is a subsidiary of Partner, which is itself a subsidiary of the Chinese firm Hutchison Telecommunications International. Pelephone was initially invested in by Shamrock Holdings (the investment arm of the Disney Group). MIRS, short for Motorola Integrated Radio Systems, was a subsidiary of Motorola-Israel, and sold in 2009 to Altice Group, a European telecommunications and media private investment/equity firm.

10. Within Israel itself, according to the National Master Plan 36A, cellular companies could erect antennas anywhere they wanted without having to notify neighborhoods and communities in advance and without providing the latter any means of objecting to these structures. The only authorities entitled to combat the antennas are the Defense Ministry and the Civil Aviation Authority, and in certain cases the Environment Ministry. This had led Israeli critics to claim that the country had become a "cellular dictatorship." Arik Merovski, "Combatting the Cellular Dictatorship," Ha'aretz, June 27, 2005, 7.

11. "The Cellular Companies and the Occupation," *WhoProfits Report*, August 2009, available at http://whoprofits.org/content/cellular-companies-and-occupation.

12. Oren Yiftachel, *Ethnocracy: Land and Identity Politics in Israel/Palestine* (Philadelphia: University of Pennsylvania Press, 2006).

13. MoC, Annual Reports for the three Israeli operators.

14. Israeli military order no. 1279: 1989.

15. Oslo 2, annex III.

16. Oslo 2, annex III, emphasis added.

17. The Oslo Accords created "temporary" administrative divisions in the West Bank: Area A in theory would be under full civil and security control by the PA; Area B would have Palestinian civil control and joint Israeli and Palestinian security control; Area C would be under full Israeli civil and security control. These areas were not contiguous and amounted to 18 percent, 20 percent, and 62 percent of the West Bank, respectively.

18. In the case of the Gaza Strip, the entirety of the area was permitted to build on after the Summer 2005 disengagement—however, not within proximity of the expanding Israeli military-defined "buffer zones" inside the strip. It is important to recognize as well that since the 2006–07 blockade, not much telecommunications equipment has been permitted into the Gaza Strip.

19. Quoted in Ilene Prusher, "Palestinians Sprint to Break Israeli Grips on Phone Lines," *Christian Science Monitor*, August 10, 1996.

20. Oslo 2, annex III, article 36, schedule 5.8. Other frequencies for wireless communications, such as those to be used by police forces or hospitals, to name but two, would also be detailed in the accords.

21. Sources: Israel MoC, Bezeq Annual Report 2011, Orange Annual Report 2010, Cellcom Annual Report 2011. On the issue of operating at a minimum amount of frequency and less than other operators, see Bill Ray, "Palestine Mobile Operator Struggles for Room to Breathe," *The Register,* October 1, 2009, available at http://www.theregister.co.uk/2009/10/01/palestine_israel (accessed October 7, 2009).

22. Oslo 2, annex III, article 36.

23. Kanafani, Hakam, "Palestinian Cellular Communications—Jawwal: Success against All Odds," Palestinian Private Sector Forum 5 (July 2004): 4–5.

24. Ibid., 5.

25. Ibid., 5.

26. Oslo 2, annex III, article 36, B.5

27. The desire to open the market to competition would be a Palestinian decision, although conditioned by Israeli acceptance to provide spectrum, an area code, permit the building of an infrastructure, and so on.

28. Kanafani, "Palestinian Cellular Communications," 4.

29. Personal interview, Jawwal, June 2005.

30. Quoted in Nicky Blackburn, "Partner, Palestinian Telco in 'Roaming' Pact," *Jerusalem Post,* November 12, 1999. At the time, Orange was the third most popular provider in Israel, with Pelephone and Cellcom having clear majority stake over the market, with whom Jawwal had no agreements to either connect to each other's phone numbers or share roaming privileges. In essence this meant that Jawwal subscribers could still not call the majority of Israeli cellular users, and vice versa.

31. Jawwal, Annual Reports.

32. Pay-per-use is possible to obtain since all cellular communications take place between a phone and particular cellular tower, which cellular companies keep track of.

33. Oslo 2, annex III, article 36, A.2a and 2b.

34. The Israeli military is an exception. Any off-duty personnel who may be in or around Qalandia can use their MIRS military-provided phones for access. MIRS—being the official Israeli military provider—does not make its service and/or phones available to any Palestinians. Phones must be obtained directly from the firm or the military.

35. Tim Cresswell and Peter Merriman, eds., *Geographies of Mobilities: Practices, Spaces, Subjects* (Burlington, Vt.: Ashgate, 2011), 4.

36. A whole range of scholars are pertinent here: Ithiel de Sola Pool, *Technologies of Freedom* (Cambridge, Mass.: Belknap, 1984); Sussman, Gerald, "Urban Congregations of Capital and Communications: Redesigning Social and Spatial Boundaries," *Social Text* 60 (1999): 35–51; Saskia Sassen, *The Global City: New York, London, Tokyo* (Princeton, N.J.: Princeton University Press, 1991); Manuel Castells, *The Rise of the Network Society* (Oxford: Blackwell, 1996); Stephen Graham and Simon Marvin, *Splintering Urbanism: Networked Infrastructures, Technological Mobilities and the Urban Condition* (New York: Routledge, 2001).

37. Tim Cresswell and Peter Merriman, eds, *Geographies of Mobilities: Practices, Spaces, Subjects* (Burlington, Vt.: Ashgate, 2011).

38. See Tawil-Souri, "Qalandia Checkpoint"; and Tawil-Souri, "Uneven Borders."

PART III

Content, Protocols, Platforms

CHAPTER 8

Protocols, Packets, and Proximity

The Materiality of Internet Routing

PAUL DOURISH

O n the corner of Wilshire Boulevard and South Grand Avenue in downtown Los Angeles stands a stark, imposing, white office tower, thirty-five stories tall, named simply "One Wilshire." It looks much like any other downtown office building, although a careful listener might notice that the muted rumble of its air conditioning, audible even over noise of the downtown traffic, seems to go beyond what might normally be expected. The building directory at the security desk in the marbled lobby begins to hint, though, at what might be unusual about this building, as the companies it lists are uniformly telecommunications providers—Verizon, Covad, Level 3 Communications, and more. For a thirty-five-story building, as it turns out, One Wilshire houses very few people, but it is nonetheless quite full. The building is given over, almost entirely, to data centers and computer server rooms, colocation facilities and network equipment, with the building's high-speed network spine as critical to its operation as its architectural supports. The nerve center of the building, and its raison d'etre, is the "meet me room" on the fourth floor, an oppressive warren of telecommunications equipment in locked cages, connected overhead by a dense, tangled mesh of bright yellow single-mode fiber-optic cables. One Wilshire's meet-me room is a major Internet connection point. It is the physical site where the digital networks of corporations like Covad and Level 3 are connected together, the point at which digital messages flow from one operator's network to another. It is where the "inter" of the "internet" happens.[1]

That the vauntedly virtual world of Internet communications is in fact grounded in physical and material realities is not, of course, a novel observation. Lisa Parks has written about spatiality and territorialization in terms of the footprints of the satellites by which global communications are made real (and distinctly local);[2] Nicole Starosielski has examined the network of undersea cables snaking across the Pacific Ocean floor and the flows of capital, expertise, and people that follow their paths;[3] Steven Graham and Simon Marvin have discussed what they call the "splintering urbanism" that comes about when different networks and infrastructures create different experiences of urban space;[4] and Kazys Varnelis has examined Los Angeles in terms of its manifestations of networked infrastructures (including, specifically, One Wilshire).[5] I, too, take the materiality of digital networks as my topic here, but my concern is somewhat different. The digital materiality that concerns me is not the materiality of the infrastructures and wires but the materiality of the digital signals that cross them. I argue that data and their protocols are also material, both in their consequences for the organization of infrastructures and in their specific manifestations as flows of electrons and signals that spread out over the wires and channels that make places like One Wilshire work. While writers like Alexander Galloway have examined the politics of network protocols, and others such as Milton Mueller have written about the institutional arrangements that bring them into being, I want here to consider protocols as material that needs to be matched with and understood in relation to the brute infrastructural materialities that we encounter in places like One Wilshire.[6]

To make these questions more concrete, I will focus here in particular on the topic of Internet routing—the protocols and mechanisms that, first, allow digital data to traverse a complex, heterogeneous, and dynamic Internet and that, second, distribute the information that allows this traversal to be accomplished. In doing so, I want to suggest a new line of inquiry for examinations of digital materiality, one that moves from a study of physical infrastructural arrangements to consider the materialities at work in the protocols, representations, models, and interactions that take place within and through those infrastructures. It is for this reason that my focus here is on the materiality of Internet *routing*, not the materiality of Internet routers or the materiality of Internet routes. That is, my concern is not with the physical infrastructure as such—the cables, the servers, the switches, the buildings, and so on—but with the processes at work.

What does it mean to think of these as material? It requires first that we adopt a methodological skepticism toward the separation of domains of practice and expertise that disciplinary and institutional boundaries typically break apart—communication infrastructures, computational platforms, protocols, and applications. It involves, instead, seeing network protocols as things that are designed to serve applications, to run on computational platforms, and to control infrastructures, bound up with and contributing to the material realization of them all. It requires too that we take a historically and geographically situated view that examines the Internet not as a Platonic ideal but as a practical and political object, one that has been shaped by many different considerations and is just one of a range of possible Internets.

In his influential book *Mechanisms: New Media and the Forensic Imagination*, Matthew Kirschenbaum draws the distinction between formal and forensic materiality as aspects of media analysis.[7] He encourages media scholars to go beyond "the event on the screen" as the object of analytic attention and to examine the specific forms of technical practice that produce those events. This is not simply a call to examine the technological and material foundations of digital experiences, although that is an important consideration, particularly from the perspective of archival studies. Rather, I take it to be a call for an examination of the *relationship* between infrastructure and experience, with attention to the processes by which digital experiences are produced; and, further, to warrant an investigation of the practices of technological design that generate these arrangements. Galloway examines the notion of protocol as a manifestation of relational power, taking his cue from the pattern of technological arrangements but in general without examining their specifics. In this study, I want to maintain the relationality that both Galloway and Kirschenbaum draw our attention to, but in a manner that, first, draws on the dual nature of protocols as mechanism and inscription à la Galloway, and, second, addresses the production of the event on the screen à la Kirschenbaum.

With these perspectives in mind, I will begin by illustrating the broad approach and then proceed by degrees through the details of internetwork routing and its materialities, before returning at the end to the broader programmatic question of how this informs a more general inquiry into the materialities of information.

The Material Analysis of Protocols

In the late 1980s and early 1990s significant research attention in computer networking was devoted to ATM (Asynchronous Transfer Mode) networking as a technology for high-speed digital communications. Unlike the TCP/IP protocols familiar to Internet users, which had been developed primarily in the academic community, ATM networking was a product of government and commercial interests, in particular the large telecommunication companies, which, in many countries, operated as government-regulated monopolies. ATM networking was standardized through the International Telecommunications Union (ITU), a UN body whose members were not individual technical experts but member countries. In this forum, technological considerations and national interests were quite explicitly bound together.

Unlike TCP/IP, ATM is not a packet-switching technology, but like TCP/IP, which divides messages into smaller, individually routed units called "packets," ATM also divides messages or message streams into small, fixed-size units known as "cells," each of which carries its own addressing and control information. A key design decision in this approach concerns how small or large these fixed-size units should be. In general, larger cell sizes make more efficient use of limited transmission capacity because they increase the ratio of payload (the digital content in each cell) to header (the fixed size control information with which each cell begins). This means that more of the bits being transmitted along a cable are bits that carry digital content. On the other hand, however, larger cells mean fewer different messages can be carried along a channel in a fixed amount of time, since larger cells take longer to transmit; smaller cells can more easily be interleaved. Similarly, in the face of this difficulty, larger cells require more buffer capacity at switching units, where they may have to be stored awaiting transmission. So, larger cells make better use of limited *transmission* capacity but smaller cells make better use of limited *switching* capacity.

Arguments about transmission and switching capacity and their relative merits are important and have economic consequences, but some of the other topics that framed the debate about cell size have a more fundamental connection to questions of existing fixed infrastructure. One of these concerns the length of the transmission cables along which ATM cells would travel. When signals travel down a wire, they have a tendency to reflect off the end of the cable and the terminating equipment, sending an "echo" back down the cable along which they have traveled. In order to reduce interference between a signal and its own echo, then, it is better if the transmission is relatively short, so that by the time the echo reaches a given point, the transmission has already ended. "Relatively" short, in this case, means "short with respect to the length of the cable"—on a longer cable, a larger message can be sent without an echo interfering with the message itself. (Echo cancellation hardware can be used to reduce this problem, but it adds significant fixed costs—an important consideration when telecommunication companies are looking for new technologies that can be implemented on their existing line infrastructure.)

Consequently, debates about the best size for a cell inevitably involve differences between groups who have largely "long" wires and those who have largely "short" ones. Telecommunications operators in large countries, such as the United States, where long-haul networks of the sort on which ATM would be deployed often cover large distances, were inclined to favor larger cell sizes, whereas those in smaller countries, and particularly those who didn't have echo cancellation installed, favored smaller ones. In the standardization process, then, the United States advocated a relatively large payload size of sixty-four bytes. France and other European countries, on the other hand, argued for much smaller size cells with a thirty-two byte payload. Each country's position incorporated the particular perspectives that one might have on questions of protocol design and efficiency in the context of their own fixed infrastructures and geographical realities. Sixty-four bytes simply made for cell sizes too "long" for small countries. A compromise position was eventually struck, and ATM cell size was fixed at forty-eight bytes of payload along with five bytes of header for a total cell size of fifty-three bytes—a size deemed equally inconvenient for all parties.

The example of the debate around ATM cell sizes troubles questions of sociotechnical analysis as they appear both in technological and in sociocultural academic circles. For the technologists, it undermines a conventional idea that while networks and technological objects are used in ways that are governed by the social, they remain themselves simply technical objects. This is what Kling et al. have called the "layer-cake model" of sociotechnical analysis—the idea that the social is something that happens "after" and "over" the technical, a consequence of material arrangements that are themselves solely technical, with the social presented as being "at the top of the network stack" by analogy with the OSI network stack, an oft-used pedagogical device.[8] Sociocultural analysts, on the other hand, are skeptical of the technological determinism at the heart of those analyses and see technological arrangements as always already social. However, in these analyses, "the technical" is rarely opened up to critical scrutiny; while technological systems are understood to be amenable to (indeed, to require) sociological analysis, the specific technological arrangements and their alternatives are rarely examined in detail. What is more, where they are, the focus is often on hardware and infrastructure. While we might laugh at the poorly informed political discussion that suggests "the Internet is a series of tubes," an understanding of internetworking as more than simply just that remains rare in sociocultural analysis;[9] protocols, representations, and their dynamics as effective media are largely unexamined. Consider a second example in which this question of dynamism and effectiveness is central. Protocols need to be designed not only to fit the sizes and properties of fixed digital infrastructures. They must also be computationally feasible—that is, fit for the computational infrastructures that will process them. When thinking about the network core, this takes on particular resonance.

Craig Partridge's book *Gigabit Networking* provides a comprehensive overview of the technical issues involved in running networks at gigabit speeds (that is, at data transmission rates of more than one billion bits per second).[10] However, at the outset of his discussion of the use of the Internet TCP/IP protocols the author finds himself presented with an odd challenge: while it is clear that network technologies can transmit data at gigabit speeds—indeed, much faster—it was not at the time universally accepted that IP-based networks could run at that speed. Skeptics argued that the protocol itself—the set of conventions and rules about how data should be formatted and how it should be processed when being sent or received—made gigabit speeds impossible. In IP networks, data is processed as "packets"—discrete chunks of data that, together with some information about where the data should be sent and how it should be interpreted, are processed independently by the network. The challenge for running IP (or any protocol) at gigabit speeds is whether a network router can process a packet before the next one arrives. As network transmission rates increase, the time available to process any given packet decreases. Advocates for alternatives to IP would argue that IP couldn't "scale"—that is, that the overhead of processing the protocol data was too great, and that IP packets could not be processed fast enough to make effective use of a gigabit network. Partridge begins his discussion, then, by laying out specific software code that can interpret IP packets and can be shown to operate fast enough to enable gigabit speeds.

The very fact of Partridge's demonstration—and, more to the point, its necessity as a starting point for his discussion—highlights some significant issues in how we think about networking technologies. These are issues that might seem self-evident to computer scientists and engineers, although their very obviousness might blind us to their importance; to others who write, talk, and think about networking technologies, though, they may seem unusual. First, it highlights the idea that, although we often talk about them as though they are the same thing, what the network can do is not the same as what the transmission lines can do—that is, a transmission line might be able to transmit data more quickly than the "network" can. Second, it highlights the fact that different network protocols can have different properties, not just in terms of their formal properties but also in terms of their practical capacities—a protocol does not simply imply rules and conventions (see Galloway) but is also subject to analyses that focus on weight and speed. Third, it draws our attention to the relationship between the "internals" of a network—the practical manifestations of how it operates—and the "externals"—that is, what it can do for us and how.

The broader, programmatic point to be made here is that a materialist concern with the Internet needs to be engaged not just with what "networks" are but rather with what this particular network—as a specific case, and as one among a range of alternatives—might be. It is not enough to argue for the critical role of decentralization, to examine the formalized disengagement afforded by protocols, or to note the geographical siting of infrastructure. Rather, I argue that a materialist account must examine just how that decentralization becomes possible, what specifically the protocols do and how, and how specific forms of spatial and institutional arrangements become possible in *just this* Internet—one that is materially, geographically, and historically specific. It is for this reason that I want to take as my focus here the question of Internet routing, as arguably one of the key conditions on what the Internet—our Internet, our current Internet—actually is, highlighting its material specificities.

Fundamentals of Internet Routing

Although we talk casually about "the Internet" or "the network," the very terms "Internet," "Internet protocol," and "internetworking" point to a more complicated reality, which is that there are multiple networks. In fact, this is the point. The crucial feature of Internet technology is that it provides a way for data to move across multiple networks, potentially of quite different sorts. The Internet links a series of networks together—local area networks on a college campus, long-distance networks operated by commercial providers, cellular networks maintained by mobile service operators, Wi-Fi networks in homes, and so on—in such a way that data can move easily from one to another.

Routing refers to the function whereby Internet messages or packets get from their source to their destination or, more accurately, from the network to which their source computer is connected to the network to which their destination computer is connected. Packets might have to traverse multiple other networks in order to get from one to the other. Those networks might be of dissimilar types, they might be owned and managed by different authorities, and there might be multiple alternative routes or paths. In this model, a "network" is an individual span of some kind of digital medium, and so it might be quite small. For instance, it is not uncommon for an average American home network to incorporate three different networks—a Wi-Fi network, a wired network (into which your Wi-Fi router is connected), and another "network" which constitutes the connection between your broadband modem and the service provider. Each transmission from your laptop to the outside world must start off by traversing these three separate but connected networks. Similarly, there are many networks inside the Internet infrastructure. For example, from where I sit writing this text in a hotel lobby in Paris, the UNIX *traceroute* utility reveals that a connection to my university's web server traverses more than twenty networks, including the local Wi-Fi network, several networks operated by international Internet provider Proxad (including networks in London, New York, and Palo Alto), several networks operated by the California nonprofit academic network operator Cenic, and finally the multiple networks of my own university. Routing is the process by which packets are correctly directed across these different network connections to reach their destinations.

Internet routing depends on three key ideas—gateways, routing tables, and network addresses. Gateways (also known as routers) are central elements in routing. A gateway is essentially a computer that is connected simultaneously to two or more networks and thus has the capacity to receive packets via one network and then retransmit them on (or "route them" to) another. For instance, a domestic broadband modem is connected both to your home network and to your service provider's network, and so it can act as a gateway that moves packets from one to the other; when it receives a message on your local network that is destined for the wider Internet, it will retransmit the message on the connection that links it to your service provider, where another router will see the message and retransmit it appropriately. Most laptops and desktop PCs have the capability to act as gateways if they are connected to more than one network simultaneously, although most gateways are actually dedicated devices (like your broadband modem or, on a larger scale, routers made by companies such as Cisco). A gateway, then, can route packets from one network onto another. To do so, though, the gateway requires information about how the network is organized.

The second key element is the information a gateway needs in order to successfully route packets. In general, this information is captured by a gateway's "routing tables," a list that associates destinations with networks. These can be thought of as rules that say, for example, "If you see a packet that is destined for X, send it to network Y." When a gateway receives a packet, it looks up these rules to determine which of its connected networks should receive it. Note that a rule of this sort does not imply that destination X is directly connected to network Y; it might simply be that another gateway connected to network Y is one step "closer" to destination X. Routing, in other words, is decentralized in TCP/IP. There is no central authority where the topology of the network is entirely known, nor any single point from which all networks can be reached. Rather, a packet makes its way across, say, the twenty "hops" from my Paris hotel lobby to UC Irvine's servers through a series of individual decisions made at each gateway it passes. A packet is passed from gateway to gateway according to routing tables that move it closer to its destination until it is finally delivered to a network to which its destination host is directly connected.

This brings us to our third concern with network addresses. Routing tables would become impossibly large if they needed to store routes for each host connected to the Internet. Instead, they record routes to networks, not to individual hosts. This requires, in turn, that some means be found to identify and name networks. A typical IP address—the familiar four-byte number like 128.195.188.233—identifies a particular host, but some part of the address is seen as numbering the network to which that host is connected. The precise mechanism by which networks are numbered will be discussed in more detail later, but for now it is sufficient to note that routing tables do not record routes for every single host address, but rather for networks (for example, 128.195.188.0, a network to which the host 128.195.188.233 might be connected.)

Before looking more directly at the protocol issues involved in distributing routing information, it is worth pausing to note some materialist concerns at work even at this foundational level. First, we should be attentive to the questions of topologies and temporalities. In small and stable internetworks, routing is a relatively straightforward operation. However, as the internetwork grows larger, the information needed to produce effective routes also grows, as does the computational power needed to process it. Similarly, in a network that is often changing, the potential paths are also highly variable. Technically, the topology of the Internet changes every time someone unplugs a cable or powers down a Wi-Fi hotspot. In a network of the scale and geographical distribution as the Internet, those sorts of changes are happening constantly. Particular solutions to the problem of routing embody assumptions about the significance, the pace, and the consequences of change and disruption.

Second, we need to be concerned as well with issues of bounds and scale. The question of routing—and in particular, its decentralized decision-making process, which we will revisit—draws attention to how particular kinds of boundaries and particular scales of operation and significance arise in the network-as-practiced. That is, the question of the temporality of changing topologies also creates zones of social, organizational, and institutional autonomy and dependence, and forces the emergence of scales and structures of control. This issue will become more important in the discussion to follow.

Third, it suggests that we might need to distinguish between protocol, implementation, and embodiment. The distinction between protocol and implementation is well recognized: we understand, analytically, the distinction between those formal descriptions of how systems interoperate on the one hand and the specific pieces of software that implement those protocols on the other—the fact that protocols are rarely complete, for example, or at least that implementers have discretion in the choices they make about the extent to which deviations from the protocol will be accepted. The distinction between protocol and embodiment speaks to a different issue. It highlights the fact that a protocol might be clear, well-defined, and effective in design and yet ineffective or inoperable in practice—when routing tables are too large, for example, or when network connections are too slow, or when routing hardware lacks the computational resources needed to process the protocol, or where the protocol is poorly matched to local conditions. A failure of protocol-connected systems is not in itself a failure of protocol, or even necessarily of implementation; specific embodiments, not just of infrastructure but crucially also of the protocol itself—data on the wire—also matter.

The fourth consideration involves decentralization, deferment, and delegation. The decentralization of Internet operations—the fact that packets are routed without appeal to a central authority, and that Internet policy is driven by what is called the "end-to-end" model, which argues for placing control at the edges of the network—is one of the most widely acknowledged features of the Internet as a specific technology.[11] However, one of the things that an examination of Internet routing reveals is that the flexibility of decentralized routing depends on many other components that may not have that same degree of decentralized control. Galloway has noted what kinds of commitments to collective agreement are implied by decentralization within a regime of protocol-driven interaction.[12] We might also point to questions of network addressing and topology as places where decentralization operates within a framework of deferment of authority and delegation to others.

Fifth and finally, an understanding of the operation and specific manifestations of routing and routing protocols needs to be seen within the context of conventions of use and practice. This point will come to be of central importance below, but it should be clear even in the discussion so far that the effectiveness of Internet routing depends not simply on the operation of the protocols but on the relationship between protocol and conventions of use—conventions that govern patterns of network addressing, for example, or topologies and practices of connectivity among service providers, or our conventional patterns of distinction between those services that are provided "close to the core" or "at the periphery" of the network. One can never rely purely on what the protocol defines or how the mechanisms operate for an account of the specifics of how our networks work.

Bearing these considerations in mind, let's proceed to another level of the analysis of routing: the protocols that govern the distribution of routing information.

Routing Protocols

Efficient and effective Internet routing depends on information about local network topology being available to routers. However, as we have seen, the decentralized nature of the Internet means that there can be no single source of information about the structure of the Internet. Routing is wayfinding without a map; it is based instead on local decisions using the best-available information. Just as routing decisions are made in a decentralized manner, the sharing of information on which those decisions are based is similarly decentralized. Routers periodically exchange information about available network connections and routes to update each other's tables. Routing protocols define how information about network topology and available routes—the information that routers store in their routing tables and use to make routing decisions—spread through the network.

There is no single, universal routing protocol. Different protocols exist for different needs, and different protocols have predominated at different historical moments. I will examine two protocols here—the Routing Information Protocol (RIP) and the Exterior Gateway Protocol (EGP).

RIP: The Routing Information Protocol

One of the earliest Internet routing protocols, and one of the most widely deployed, is RIP: the Routing Information Protocol. RIP's widespread adoption in the 1980s and 1990s derived not from its technical superiority but from the fact that it was implemented by the *routed* (pronounced "route-dee") software distributed with the 4BSD Unix software distribution, popular in academic environments, and later with Sun Microsystem's SunOS operating system. In fact, this software was for some time the only available specification of the protocol: there was no formal description or definition, merely the behavior of the implementation. Not only was RIP an early protocol for exchanging Internet routing information, but it was heir to a longer legacy; RIP is a variant of the routing information protocol that formed part of Xerox's Network Service protocol suite (XNS), which itself embodied ideas originally developed as part of Xerox's PUP and Grapevine systems.[13]

A router running RIP will periodically broadcast a description of its routing tables to other routers that it knows about. The information that an RIP router provides is essentially its own perspective on the network—that is, it describes everything relative to itself. RIP provides two pieces of information about each of the routes—its destination and the hop count. Hop count—the number of routers or network segments to be traversed in order to reach the destination—serves as an approximate metric of distance. Some networks might be fast, some slow; some might cover long distances, some shorter ones. These distinctions are not, however, captured in the hop count, which provides a more rough-and-ready basis for decision making about the most efficient route.

RIP uses just four bits to record the hop count, allowing it to indicate a range of values from zero to fifteen. A value of fifteen indicates an "infinite" hop count, used to signal that a network is unreachable. Accordingly, a network in which routes are communicated solely via RIP can be no more than fifteen networks "wide"—in other words, a packet cannot be routed across more than fifteen segments unless other measures are taken. In practice, this makes RIP useful primarily in smaller networks that are themselves connected using different routing protocols; the global reach of "the Internet," in other words, is premised on such specificities.

EGP: The Exterior Gateway Protocol

EGP, the Exterior Gateway Protocol, is a more complex protocol than RIP. Itself an evolution of an earlier protocol called GGP, it is designed for communication between routers that connect so-called "autonomous systems"—networks that span particular organizations, corporations, or institutions. Intuitively, if you imagine that a university such as UC Irvine or a corporation such as Intel each runs its own networks according to the institution's own conventions and pro-cedures, then each is designated as an autonomous system; EGP is the protocol by which routing information about one of these networks is communicated to routers for the other.

As with RIP, the core of the EGP is a mechanism by which routes are shared from one router to another. Also like RIP, hop count is used as a rough-and-ready measure of distance, although unlike RIP, routes of more than fifteen hops can be expressed. EGP expresses the "distance" to particular destinations relative to specific, identified gateways (rather than implicitly from the sending router). The protocol is also more fully featured than that of RIP; for instance, there is an explicit component by which new neighboring gateways can be identified and polled. By contrast, this structure in RIP is left as a matter of configuration.

The intended purpose and conventional use of EGP differs from those of RIP, which is not committed to particular forms of use, although its constraints limit the conditions under which it can be deployed. EGP, on the other hand, is designed specifically to connect autonomous systems. Accordingly, EGP is designed to be used with particular conventions regarding which routes should be advertised and which should not; these conventions are not encoded directly in the protocol, but rather the protocol is designed under the assumption that administrators will be conscious of them.

Material Considerations

Network protocols are shaped by material constraints. ATM cells have not just an abstract size but also a length when transmitted along cables. IP packets do not simply have a format, they have a format that has consequences for the speed of processing in network routers and so can be limited by switch fabrics. Similarly, the centrality of routing to the Internet can be understood materially in terms of the arrangement of network nodes, the cost of routing, the structure of networks, the size of routing tables, and the dynamics of connectivity. Critically, this materiality cuts across apparently different domains of concern—from the practice of network operations to the rhetoric of democratic access. I will consider four aspects here.

Routing Tables, Classless Routing, and the Politics of Address Exhaustion

In 1993, changes were introduced to the way that Internet routing worked. The new model—called CIDR (pronounced "cider"), or Classless Inter-Domain Routing—extended and built upon conventions of "subnet routing" that had been in operation for some time but were adopted as the basis of a new model of routing that would apply across the Internet.[14] CIDR was a response to a growing problem for Internet routers, particularly core routers: the size of routing tables. The large number of networks meant that routing tables were growing, with three consequences: the storage demands on each router were growing significantly, beyond what had ever been imagined; the processing time necessary to sort through the routes was also growing; and the process of transmitting the routing tables (via protocols like EGP) was becoming unwieldy because the routing tables were so large.

CIDR was known as "classless" routing because it replaced an earlier scheme of network addressing that distinguished between three "classes" of networks, A, B, and C. Class A network addresses were distinguished by just their first eight bits, so that, for instance, all addresses that begin 13.x.x.x belong to Xerox Corporation. Holding a class A network allocation meant that one had control of a total of 16,777,216 separate Internet addresses (although a handful are reserved and cannot be used.) Class B addresses were distinguished by their first two bytes (sixteen bits). Addresses in the range 192.168.x.x, for example, are B addresses designated for private use. Each class B allocation includes 65,536 addresses (with the same caveat.) Class C network addresses were distinguished by their first three bytes and provide 256 addresses (again, minus a handful).

In the original scheme—which, by contrast with classless routing became known as "classful" routing—the three classes of addresses served two simultaneous functions. They were the units of routing because, as discussed above, routes direct packets toward networks rather than to individual hosts; and, at the same time, they were also the units of allocation. If a new entity needed some address space—perhaps a university, a government, a company, or an ISP—then it would be allocated one or more class A, class B, or class C addresses. This conflation of two functions then resulted in two interrelated problems that classless routing could, it was hoped, solve. The technical problem was the growth of routing tables; the political problem was the growing scarcity of address space.

In classful addressing, networks are either class A, class B, or class C; the division between the network portion of an Internet address and the host portion occurs at an eight-bit boundary, so that in a class A address the thirty-two bits of an IP address are divided into eight bits of network identifier and twenty-four bits of host, rather than sixteen and sixteen for class B, and twenty-four and eight for class C. Classless addressing introduced a mechanism whereby the boundary between network and host portions could be made more flexible.

This mechanism would address the technical problem, the growth of routing tables, by allowing smaller networks to be coalesced into a single network that was still smaller than the next class up. This was especially important for class C addresses. A class C address covers only around 250 possible hosts. For many organizations that didn't need a whole class B address, a class C network was too small. So many organizations would be allocated many class C network addresses—each of which would require individual entries in routing tables. By creating the opportunity to have, say, ten bits of host space (for around twelve hundred hosts) rather than eight bits—a new arrangement not possible in traditional classful routing—classless routing could shrink the size of the routing tables by dealing with networks at a new level of granularity.

This also addressed, to some extent, the political problem. The introduction of classless routing may have been sparked not least by the troublesome growth of routing tables, but it directly addresses another regular concern around Internet management and governance, the question of address allocation and exhaustion. The original strategy for Internet address allocation was based on address classes, with class A addresses particularly prized among large organizations. This resulted in various well-known inequities of allocation, such as the fact that MIT (with the class A network address now known as 18.0.0.0/8) has more allocated addresses than China.

Classful address allocation suffers a second problem, which is address wastage. Xerox, for instance, was allocated the class A network 13.0.0.0, although it seems unlikely that they would use all 16 million available addresses; however, at around sixty-five thousand addresses, the next step down (class B) was deemed likely to small. No effective mechanism was available for smaller allocations. It remains the case, even in a world of classless routing, that the IP address space is continually approaching exhaustion, even as we know that wastage goes on within the already allocated blocks.[15] Again, this is also happening within the context of deep historical inequities in address allocation, as noted above.

The fact that routing presents both technical problems and political problems is not surprising. What is important here is the material entwining of these problems—the fact that the politics of network-address space allocation and the dynamics of routing-table growth and exchange are dual aspects of the same material configurations. The political and technical issues are not so much twin problems as different facets of the same problem, which is that in the interests of efficiency and effectiveness, networks are the primary units of route determination. When we see these in a material light—that is, when we see route advertisements as things that have size, weight, lifetimes, and dynamics, then the problems become material too.

Granularity and Networks as User Objects

This discussion has been based on a set of technical conditions that govern what "a network" is, in Internet terms—not an abstract large-scale entity ("the Internet" rather than the ATM network) or an autonomous system ("UC Irvine's network") or even the entities of common experience ("my home network"), because none of these are the sorts of "networks" of which the Internet is a connective fabric. Rather, the "networks" that the Internet connects are particular media arrangements—lengths of coaxial or fiber-optic cable, wireless signal fields, tangles of twisted-pair wires, and so on. These networks are technical entities but not user entities.

Or, at least, not usually. The vagaries of routing can also result in the infrastructure of network arrangements becoming visible and even important as a question of user interaction. Voida et al. document an interesting and unexpected case—the case of music sharing within a corporate network.[16] A network in IP is not only the unit of routing, but also the unit of broadcast—that is, messages can be delivered to all the hosts on a particular network. This in turn means that networks can become visible as marking the boundaries of broadcast-based sharing. In the case that Voida and colleagues document, participants in a corporate network come to realize aspects of its internal structure—that is, the way that it is composed of multiple subnetworks connected via routers—through the patterns of visibility and invisibility that they can perceive through the use of music sharing in the iTunes application. iTunes can allow people on a network to share music with each other, and the corporate participants in the study took advantage of this facility. However, they began to encounter problems that resulted from the network topology—that they couldn't see a friend's music, for example, because the friend was on a different network, or that someone's music might "disappear" if that person relocated to a different office, even within the same building. The network topology itself became visible as a routine part of their interactions, and suddenly the material arrangements that underlay the notion of "the network" became an aspect of the user experience.

What is significant here is the collapse of the network stack—the tower of abstractions that separates the physical medium from internetworking protocols, and internetworking from applications. Suddenly, in this model, the specific material manifestation of IP networks—as runs of cable governed by signal degradation over distance and the number of available sockets on a switch—need to be managed not only for the network but also for the users. Network engineers and researchers have long recognized that the abstractions of the network stack may be a good way to talk about and think about networks but a less effective way of managing or building them.[17] However, the intrusion of the material fabric of the network and its routing protocols into the user experience is a different level of concern. If granularity is a core material property, one that shapes the decomposition of the network into effective units of control, then Voida and colleagues highlight the way that this materiality manifests itself in multiple different regimes.

The Emergence of Centralized Structure

The mythology of the Internet holds that it was designed to withstand nuclear assault during the Cold War era, both by avoiding centralized control and by using a packet-switching model that could "route around" damage, seeking new routes if any node were lost. Whatever the status of that claim,[18] it is certainly true that one of the defining features of the Internet is its variable and amorphous topology. Unlike networks that rely on a fixed arrangement of nodes (such as a ring-formation in, for example, the Cambridge Ring), the Internet allows connections to be formed between any two points, resulting in a loosely structured pattern of interconnections (what computer scientists call "a graph").[19] The absence of formal structure and the avoidance of critical "choke points" or centralized points of control is arguably one of the essential characters of the Internet.

However, our examination of routing and routing protocols reveals a more complicated situation at work. The contrast between the operation and the operating context of RIP and EGP is educational in this respect. RIP is a simple protocol that predates the Internet; EGP, on the other hand, is a protocol that emerged over time and evolved in response to the conventional practices of the Internet as a set of practical institutional arrangements. As described, EGP is based around the idea of autonomous systems—the idea that different networks will belong to different institutional entities and will be managed autonomously by different authorities. It is also based on the idea that access to each autonomous system will be brokered by one or a small number of authoritative gateways. The conventions that govern the use of EGP—for instance, the rule that no gateway may advertise a route to a network other than those within the autonomous system it represents—foster this concentration of authority. In other words, what we see, within the framework of the open connection, open routing, and amorphous structure afforded by the Internet's fundamental technologies, is the emergence of authority, control, institutional structure, and local points of centralization. Centralization may not be inscribed in the basic protocols of TCP/IP but may emerge at other points as a consequence of practicality.[20]

This raises some interesting questions. One is: Precisely which Internet do we talk about when we celebrate openness, diversity, and decentralization as characteristics of the Internet when compared to mass media as forms of communication? Certainly, we can celebrate the potential for these properties, but perhaps not their practical embodiment within the Internet as we know it—our Internet rather than a possible Internet. It is not at all clear that the Internet, our Internet, is in fact the decentralized, open, and democratic tool of connection and communication that technolibertarian rhetoric applauds. Second, it is important to see the kinds of centralizing tendencies and emerging structure and conventions that EGP represents and encodes as material consequences of the Internet's form—the dynamics of its topological organization, the pragmatics of routing, the consequences of bandwidth provision, the economics of access, and so forth. That is, they are not consequences of Machiavellian dabbling, of corporate subversion, of capitalist corruption, or of state intrusion. In the spirit of Kelty's recursive publics, these are protocols and conventions, after all, that have been developed by the very people who hold dear the Internet's independence from these constraints.[21] Rather, as I have attempted to show here, they are material consequences of the relationship between infrastructure and protocol, between representation and practice, and between encoding and practical action.

Historical Patternings

As we have seen, the routing protocols implemented on the Internet reflect a historical pattern of evolution. EGP grew out of GGP; it was itself superseded by BGP (the Border Gateway Protocol), which implements CIDR and has been in use since the mid-1990s. RIP was derived from earlier protocols developed at Xerox—first PUP and then XNS; the XNS routing protocol similarly became the basis for routing in Novell's NetWare product suite. From one protocol generation to the next, certain ideas and expectations are inherited in the technical design. What else is inherited along with those technical features? Each protocol is designed to capture both what has worked well in a protocol that came before it and to correct or respond to problems that have arisen. Assessments of success and failure, and the identification of effective and ineffective properties, are made relative not to designs but to deployments.

One of the central considerations that arises when we see the protocols as emerging out of deployments rather than simply as technical designs is the issue of control, authority, and management. The question essentially becomes to what extent the network should operate as a self-organizing, adaptive entity (which is the principle embedded in the routing algorithms themselves) and to what extent it is an entity that is actively managed (the principle increasingly embedded in the protocols by which routing information is distributed). In the evolution of GGP to EGP to BGP, we see similarly an evolution in understandings of the degree of control and management needed within core networking routing. Within the history of the protocols of which RIP was a part, we also see some important considerations in the expectations of deployment. PUP and XNS were protocols deployed within formal organizations (PUP internal to Xerox, and later XNS as a product for Xerox's corporate customers); similarly, Novell's later NetWare product was also a product for corporate networking. Corporate network management generally implies the presence of network administrators and a policy function to manage how networks are designed, deployed, and operated. What we find embedded in the protocols are organizational expectations about structure and management—the constraints within which the flexibility of an adaptive, evolving, self-managing network can operate. Indeed, the question we might want to ask at each juncture where the open and self-managed nature of the network appears, such as in routing, is: What structures or constraints are needed to allow this flexibility?

Conclusion

The mathematical computer scientist Edsger Dijkstra is reputed to have remarked, "Computer science is no more about computers than astronomy is about telescopes."[22] We may read this as a comment that computation should be seen as independent of its material manifestations. By contrast, recent interest in "information infrastructure studies" has demonstrated the importance of turning attention to the infrastructures of contemporary information systems in order to examine the processes and conditions under which information and information systems are produced, maintained, and put to work.[23] Whether or not this is a consideration for computer science as a discipline, it is most certainly an important consideration for the way computer science and its products manifest themselves in our everyday world—and for the way that computer science as a discipline evolves and develops.

In framing this article, I distinguished between the materialities of Internet routes, of Internet routers, and of Internet routing as distinct topics of investigation. Of course, at places like One Wilshire, these come together. In July 2013 the One Wilshire building was sold for more than $430 million—the highest price per square foot ever commanded for real estate in downtown Los Angeles.[24] The materialities of the routes signaled by the spray-painted markings on the street outside the building (tracing conduits below) and the materialities of the routers and devices the building hosts, powers, and cools are important ways into understanding the realities of contemporary digital life, but the protocols that tie these things together—that make the conduits effective, that enliven the servers, that allow them to operate productively—must also be part of the picture.

This chapter arises as part of a larger investigation of what colleagues and I have been calling the "materialities of information."[25] The twin foundations of this project are, first, that recent interests in materiality arising in the social sciences and humanities provide an important basis for understanding contemporary technological phenomena, with an attentiveness not just toward infrastructure but toward information itself;[26] and, second, that to do so effectively requires a foundational engagement with the computational objects and processes that make up the technological landscape.

Our interest in materiality is not taxonomic—that is, our goal is not to redraw the boundaries that separate "the material" from "the immaterial." Our concern instead is to examine the material considerations within the body of technical and social practice that constitutes the contemporary regime of information. Internet routes and Internet routers have been productively examined from the perspective of materiality;[27] however, turning a materialist eye upon Internet routing reveals the entangling of protocol, politics, and pragmatics that come together not only at physical sites like One Wilshire but in the materialization of protocols like EGP as embedded within systems of practice and technological artifacts. Indeed, I would argue that an examination of the materialities of information must engage with information systems not simply as metaphors of virtuality but as historically and geographically specific configurations of technology and practice. This provides an opportunity to frame an investigation of the materialities of information as what Pickering has called a "real-time understanding."[28] Routing—as manifested in *our* Internet, in *the* Internet, in *this* Internet, rather than in *an* Internet—provides an example of doing so.

Notes

1. The explorations of materiality presented here have arisen primarily in conversation with Melissa Mazmanian, whose contributions are central. Nicole Starosielski provided thoughtful and useful comments on an earlier draft. This work is supported in part by the National Science Foundation under awards 0917401, 0968616, and 1025761, and by the Intel Science and Technology Center for Social Computing.

2. Lisa Parks, "Satellites, Oil, and Footprints: Eutelsat, Kazsat, and Post-Communist Territories in Central Asia," in *Down to Earth: Satellite Technologies, Industries, and Cultures*, ed. Lisa Parks and James Schwoch (New Brunswick: Rutgers University Press, 2012).

3. Nicole Starosielski, "Beaches, Fields, and other Network Environments," *Octopus Journal* 5 (2011): 1–7.

4. Steven Graham and Simon Marvin, *Splintering Urbanism: Networked Infrastructures, Technological Mobilities and the Urban Condition* (London: Routledge, 2001).

5. Kazys Varnelis, *The Infrastructural City: Networked Ecologies in Los Angeles* (Barcelona: Actar, 2008).

6. Alexander Galloway, *Protocol: How Control Exists after Decentralization* (Cambridge, Mass.: MIT Press, 2004); Milton Mueller, *Networks and States: The Global Politics of Internet Governance* (Cambridge, MA: MIT Press, 2010).

7. Matthew Kirschenbaum, *Mechanisms: New Media and the Forensic Imagination* (Cambridge, Mass.: MIT Press, 2008).

8. Rob Kling, G. McKim, J. Fortuna, and A. King, "Scientific Collaborations as Socio-Technical Interaction Networks: A Theoretical Perspective," *Proceedings of the American Conference on Information Systems*, Long Beach, California, 2000; Andrew Tanenbaum and Davis Wetherall, *Computer Networks*, 5th ed. (Boston: Prentice-Hall, 2010).

9. Ken Belson, "Senator's Slip of the Tongue Keeps on Truckin' over the Web," *New York Times*, July 17, 2006.

10. Craig Partridge, *Gigabit Networking* (Boston: Addison-Wesley, 1994).

11. J. H. Saltzer, D. P. Reed, and D. D. Clark, "End-to-End Arguments in System Design," *ACM Transactions on Computer Systems* 2, no. 4 (1984): 277–88; Tarleton Gillespie, "Engineering a Principle: 'End-to-End' in the Design of the Internet," *Social Studies of Science* 36, no. 3 (2006): 427–57.

12. Galloway, *Protocol*.

13. D. Oppen and Y. Dalal, "The Clearinghouse: A Decentralized Agent for Locating Named Objects in a Distributed Environment," *Office Systems Division Tech Report OSD-T8103* (Palo Alto, Calif.: Xerox Corporation), 1981; A. Birrell, R. Levin, R. Needham, and M. Schroeder, "Grapevine: An Exercise in Distributed Computing," *Communications of the ACM*, 25, no. 4 (1982): 260–74; D. Boggs, J. Shoch, E. Taft, and R. Metcalfe, "Pup: An Internetwork Architecture," *IEEE Transactions on Communications* 28, no. 4 (1980): 612–24.

14. V. Fuller, T. Li, J. Yu, and K. Varadhan, "Classless Inter-Domain Routing (CIDR): An Address Assignment and Aggregation Strategy," *RFC 1519*, Internet Engineering Task Force, 1993.

15. Z. Wang and J. Crowcroft, "A Two-Tier Address Structure for the Internet: A Solution to the Problem of Address Space Exhaustion," *RFC 1335*, Internet Engineering Task Force, 1992; Mueller, *Networks and States* (Cambridge, Mass.: MIT Press, 2010).

16. A. Voida, R. Grinter, N. Duchenaut, K. Edwards, and M. Newman, "Listening In: Practices Surrounding iTunes Music Sharing," *Proceedings from the ACM Conf. Human Factors in Computing Systems CHI 2005*, Portland, Oregon, 2005, 191–200.

17. D. Clark and D. Tennenhouse, "Architectural Considerations for a New Generation of Protocols," *ACM SIGCOMM Communications Review* 20, no. 4 (1990) 200–208.

18. It is true that Paul Baran's original report on packet switching was inspired by the nuclear-assault scenario, and that the developers of the Internet recognized the originality and value of this approach to network design; however, the extent to which the Internet or its ARPANET predecessor was designed with this goal in mind is highly questionable. Paul Baran, "On Distributed Communications," *RAND Memorandum RM-4320-PR* (Santa Monica, Calif.: Rand Corp., 1964).

19. A. Hopper and R. Needham, "The Cambridge Fast-Ring Networking System," *IEEE Transactions on Computers* 37, no. 10 (1988) 1214–23.

20. EGP was replaced by BGP, which has different properties. Unlike EGP, which runs only at the edges of autonomous systems, BGP is also used internally, which provides some opportunities to defuse the pattern of centralization that arises with EGP. However, in BGP, routing is governed by explicit policies rather than by distance metrics, representing to some degree an assertion of authority and control over autonomy and adaptation. The same tendencies, then, remain at work.

21. Chris Kelty, *Two Bits: The Cultural Significance of Free Software* (Durham, N.C.: Duke University Press, 2008).

22. It is at best questionable whether Dijkstra himself ever said this; rarely at a loss for a pithy comment, Dijkstra, like Mark Twain or Winston Churchill, was one of those quotable characters to whom all manner of comments are often attributed.

23. Geoffrey Bowker, Karen Baker, Florence Millerand, and David Ribes, "Toward Information Infrastructure Studies: Ways of Knowing in a Networked Environment" in *The International Handbook of Internet Research*, ed. Jeremy Hunsinger, Lisbeth Klastrup, and Matthew M. Allen (New York: Springer, 2010), 97–117.

24. Roger Vincent, "Downtown L.A. Office Building Sells for Record $437.5 Million," *Los Angeles Times*, July 17, 2013.

25. Paul Dourish and Melissa Mazmanian, "Media as Material: Information Representations as Material Foundations for Organizational Practice," in *How Matter Matters: Objects, Artifacts, and Materiality in Organization Studies,* ed. Paul Carlile, David Nicolini, Ann Langley, and Haridimous Tsoukas (Oxford University Press, 2013), 92–118.

26. Daniel Miller, ed., *Materiality* (Durham, N.C.: Duke University Press, 2005); Matthew Kirschenbaum, *Mechanisms: New Media and the Forensic Imagination* (Cambridge, Mass.: MIT Press, 2008); Wanda Orlikowski, and Susan Scott, "Sociomateriality: Challenging the Separation of Technology, Work and Organization," *Academy of Management Annals* 2, no. 1 (2008): 433–74.

27. Adrian Mackenzie, "Untangling the Unwired: Wi-Fi and the Cultural Inversion of Infrastructure," *Space and Culture* 8, no. 3 (2005): 269–85; Kazys Varnelis, *The Infrastructural City: Networked Ecologies in Los Angeles* (Barcelona: Actar, 2008); Starosielski, "Beaches."

28. Andrew Pickering, *The Mangle of Practice: Time, Agency, and Science* (Chicago: University of Chicago Press, 1995).

CHAPTER 9

Service Providers as Digital Media Infrastructure

Turkey's Cybercafé Operators

SARAH HARRIS

While conducting field research at a cybercafé in Erzurum, Turkey, in 2010, my attention shifted from the computer to the garbled sounds of walkie-talkie static. A stern voice interrupted the hum of mouse clicks, keyboard tapping, and hushed online conversations. Through a narrow opening in the curtain dividing the male section in the front of the café from the female section in the back, I could see a police officer checking each customer's ID to confirm that he met the minimum age requirement of eighteen. I wondered if the cybercafé operator, who was required by law to filter and monitor his customers' online activities, would alert the officer to my search for proxy servers that bypassed the national Internet filters or to my neighbors' use of similar strategies to access YouTube (which was banned at that time). To my relief, the operator chose to keep these circumventions private, likely to avoid fines and closure, while the officer completed his check of the male customers, never entering the women's section, and left. The cybercafé operator had thus served a mediating function between Internet users and state authority in two ways. Not only had he chosen not to report our illegal activities, but by partitioning his café into male and female sections, he had extended Internet access—and the possibility to circumvent website bans via proxies—to women from pious, conservative backgrounds, who wouldn't normally visit Erzurum's coed cybercafés used mainly by men.

As Internet use rose in Turkey from approximately 4 percent in 2000 to 45 percent by 2012,[1] cybercafés were increasingly scapegoated for an array of societal ills and targeted by state policies and law enforcement. Erzurum cyber-café operators pointed out that because locals are deeply pious, police visits were especially frequent. Yet their visits were often surface-level sweeps, performed to appease the older generations alarmed by press scrutiny of cybercafés. Delicately hiding their customers' banal circumventions while acquiescing to the police, the operators were in a paradoxical position: despite being designated enforcers of filtering and surveillance at the endpoints of Turkey's Internet infrastructure, operators also help their customers bypass state and corporate restrictions on banned or proprietary content. Their choices—to disclose or hide surveillance data, to cooperate with or resist filtering mandates, to create gender-inclusive or exclusive cafés, and to pirate or purchase licenses—are as important to ICT infrastructure as the availability of a functioning, material base.

In this chapter I explore how the technological, social, and regulatory practices of cybercafé operators have shaped ICT infrastructure in Turkey. I approach Turkey's cybercafés as nodes of ICT infrastructure where national policies and circumvention practices are negotiated, and where access gaps are addressed. In the first section I examine how cybercafé operators have unionized in response to a national concern that the Internet is a harbinger of moral indecency and social unrest. The Istanbul Internet Café Operators Union leverages the cybercafé's position as an infrastructural node to expand operators' entrepreneurial autonomy and to vie for control over national Internet filtering. The second section describes how, in spite of the risks, nonunionized operators bypass ICT laws by allowing or encouraging proxy use and piracy in order to grow their customer base. Their choices generate working-class ICT participation where it would otherwise not exist. The final section analyzes the mutability of ICT infrastructure by highlighting how operators' attitudes and design choices shape who their customers are and how they engage with technology. I focus on a small subsection of operators who promote female participation by gender-segregating their cafés to work around social protocol limiting women's access. Ultimately, this study reveals how human labor practices at cybercafés are as essential a component of ICT infrastructure as regulations, industries, and material conduits.

This research is drawn from fieldwork conducted in Turkey between 2008 and 2013, interviewing more than fifty operators and visiting cybercafés in ten cities across the country.[2] I selected cafés in neighborhoods catering to both pious and secular customers, where I investigated the labor of service provision and the relationship between technological conduits and organized social practices.[3] My research revealed vast differences in ICT access, literacy, and agency, differences that contradict the populist and cosmopolitan promises of governmental Information Society development initiatives, which began in the early 1990s, when the country's Internet was first established.[4] Utilizing a range of technical and interpersonal skills, cybercafé operators provide ICT access to users whom Jack Qiu, in his research on China's working-class network society, astutely categorized as the "information have-less" located on neither side of the digital divide but rather at the boundary between "haves" and "have-nots."[5] Cybercafés are essential access points for Turkey's information have-less—an estimated 25 percent of the total population of Internet users in Turkey who connect via Public Internet Access Points (PIAPs)[6]—and they provide participatory opportunities for the approximately 50 percent of Turkish citizens who do not yet access the Internet.

My study of cybercafé operators is particularly resonant with Greg Downey's analysis of messenger boys who occupied key positions in early telegraphy networks "at the boundary between the virtual and the physical."[7] Despite being self-taught engineers and entrepreneurs who learned their trade through

Figure 9.1. "World-Net Internet Café" advertises high-speed internet, food, drinks, faxing, mobile-phone minutes, and homework checks conducted by the operator. Gaziantep, Turkey, 2012. Photo by author.

apprenticeships, professional associations, technological repair, and online study, Turkey's cybercafé operators are—as the messenger boys once were—generally regarded as unskilled labor. Much like the messenger boys were only noticed when they were blamed for lags in telegraph speed, cybercafé operators only become visible when scapegoated for cybercrime and piracy—forms of regulatory lag. Nevertheless, these individuals play a vital role in organizing and sustaining PIAPs, in disrupting authoritarian control, and in coordinating and extending ICT infrastructure for the have-less.

Cybercafé Unions

In the late 1990s, Turkey's cybercafés were places where national filters, software licenses, and surveillance measures were haphazardly implemented, depending on the operator's preferences and resource capacity. By 2007 the rise in Internet use and proliferation of cybercafés caught the attention of the ruling conservative political party—the AKP (*Adalet Kalkinma Partisi* or "Justice and Development Party")—who pushed through legislation mandating the filtering and monitoring of PIAPs and gradually restricted Internet use in private access points as well. Through their influence over regulation and the telecommunications monopoly, the AKP government has facilitated over forty thousand website blocks to date.[8] With allies and family members in media ownership positions who publicized their cause, the government targets the cybercafés, portraying them as portals for obscenity, violence, and terrorism.[9] This scapegoating is symptomatic of operators' threat, as they provide users with proxy access and the skills needed to send and receive politically sensitive, censored content, and to pirate digital assets.

On May 4, 2007, Turkey's first Internet policy—Law 5651, "On Regulating Broadcasting on the Internet and Fighting Against Crimes Committed through Internet Broadcasting"—placed the nation's courts in charge of regulating online content.[10] According to the law, citizens, prosecutors, and officials from the Information Technology and Communication Authority, or the BTK (Bilgi Teknolojileri ve Iletişim Kurumu),[11] could bring lawsuits against websites on a case-by-case basis. If the website in question included pornography or obscenity, disparaged the nation, promoted narcotics or terrorism, or included slander or material harmful to children, a judge could suspend access. Known as "5651," the law has led to bans on websites such as YouTube (2007–2010, 2014),[12] the Kurdish news portal Firat News (2006-present), the blog of evolutionary scientist Richard Dawkins (2007–2008), the blogging platform Blogger.com (2008), and the video-sharing site Vimeo (2014). Nevertheless, for years the issue of cybercensorship remained off the radar of upper- and middle-class Internet users for two primary reasons. First, because bans were appealable, they were impermanent; as lawyers successfully appealed the YouTube and Dawkins bans, it appeared that the courts could and would undo bans that infringed on the freedom of online expression. Second, and more important, Internet filters were inefficiently applied by Türk Telekom (TT), the formerly state-owned telecommunications company that was privatized under E.U. stipulations in 2005 yet still controlled 95 percent of the backbone.[13] TT's bureaucratic administration failed to filter illegal websites from subscribers, creating a mysterious experience in 2008–09 when I could intermittently access banned content at home without the aid of a proxy.

Filtering inconsistency encouraged public apathy toward cybercensorship until a 2011 regulation shifted oversight from the courts to the BTK's "Internet Council." Composed of seven state and four civil society "experts," the Internet Council was made responsible for writing and updating website filter blacklists. Domestic ISPs who rented access from TT would become responsible for filtering the classified blacklists for their subscribers. The 2011 changes signaled an era of more extensive, consistent, and efficient cybercensorship in private access points. In response, in May 2011 tens of thousands of citizens organized public demonstrations under the name Don't Touch My Internet! (DTMI!). Bringing together diverse participants who had never before engaged in public protest, DTMI! marked the emergence of cybercensorship as a central political issue. This issue would reemerge in subsequent demonstrations, including "Gezi Park" (2013) and the 2014 protests denouncing an amendment that would further consolidate government control over domestic ISPs, enhance their ability to filter and monitor subscribers, and mandate the maintenance of surveillance logs for two years.[14]

In contrast to the late arrival of laws regulating private access, laws addressing public access were initiated earlier. The 2007 "Regulation on the Providers of Public Use Internet" was intended for the most popular public provider at that time—the cybercafé—requiring that operators enforce a limit on age, ban smoking, curtail hours of operation, use security cameras, implement state-approved filtering and surveillance software, and maintain user logs for one year (figure 9.2).[15] Alarmed by the law, some operators established the Istanbul Internet Café Operators Union or IIKO (Istanbul Internet Kafeciler Esnaf Odasi) in 2007 and similar unions in Ankara and Diyarbakir. IIKO's sixteen hundred members are cybercafé operators from working-class backgrounds, ranging from age twenty to fifty. Most have completed middle school or high school and own from one to three cafés.[16] When two IIKO organizers, Çetin and Emre, arrived for our 2011 interview at a bustling coffeehouse inside one of Istanbul's giant malls,[17] their dark suits and serious expressions starkly contrasted with the carefree, florescent surroundings. Compared with the other customers—smartphone-toting teenagers, college students on laptops, mothers with small children, and Arab and European tourists—this pensive pair appeared out of place. As I was realizing through my research, their demeanors matched the tense environments at their cybercafés, which had been targeted in ways that mall-cafés had not.

Çetin and Emre blamed cybercafé operators' desperate economic situation on government bureaucrats—whom they described as "technologically inept and ignorant of regulatory issues"—and on expensive proprietary software licenses. As Turkey's rate of home and mobile Internet use boomed around 2010,[18] operators struggled to keep their businesses afloat, searching for ways to innovate and market their services amid increasing competition and a legally punitive milieu. They explained IIKO's objectives: to amend laws that unfairly target cybercafés; to increase government support for operators to afford proprietary software; to position cybercafés as a more efficient Internet content regulator than the Internet Council; and to improve the cybercafé's stained public image. As Çetin put it, "First and foremost, we are asking for a decrease in the fines. But we also want a public acknowledgement from the government

Figure 9.2. A typical cybercafé operator's monitor through which he logs each customer's activities. Turkey, 2011. Photo by author.

that operators are people who help make the Internet safer. We shouldn't be treated like criminals."

Emre explained how the 2007 PIAP law unfairly pressured cybercafé operators while ignoring circumvention occurring at other PIAPs:

> In the last one and a half years (2009–2011), cybercafés in Istanbul have been raided over 150 times, and the fines operators face if they make a minor mistake are so excessive, it will put you out of business. If you are missing a software license, you can be fined up to 150,000 Lira ($80,500) and jailed up to five years. For missing user logs, the fine is in the tens of thousands of Lira. Now, look around the coffeehouse where we're sitting . . . Does this café face fines and negative publicity for not filtering properly? No. The entire burden for filtering and logging falls on us.[19]

The union was equally concerned with high-cost operating system (OS) licenses, namely Microsoft Windows, the most widely used OS in Turkey,[20] and have demanded greater governmental support for operators to purchase Windows. Law 5846, "The Law for Significant Ideas and Artistic Works,"[21] prohibits Turkish citizens from using, copying, or distributing copyrighted software and digital content without permission from the author.[22] Although home users can illegally access proprietary content through peer-to-peer file sharing, cybercafé operators are expected to purchase licenses, and they risk fines and imprisonment if they refuse.[23] This makes piracy both tempting and risky for the operator, who must choose between paying for licenses and paying piracy fines. Already scapegoated, IIKO representatives hesitated to acknowledge that piracy is prevalent in Turkey's cybercafés, and instead they blamed software corporations. On their website, IIKO's president singled out Microsoft, accusing the company of tricking the operators with confusing contract language and discriminatory pricing. In addition to foreign industry, the president's remarks disparaged local Internet activists—arguing that their promotion of Internet freedom and decentralized regulation had created an environment wherein regulators, fearful of being called cybercensors, would relax regulations in ways that weaken national security.[24]

Interestingly, IIKO's critiques are not directed at the government's scapegoating of cybercafés—in part because the AKP would make a formidable enemy, but also because IIKO and the AKP each promote centralized, authoritarian control. Rather than change the regulatory system altogether, IIKO is trying to shift authority from the government to themselves, from Islamist interests to Kemalist ones, and from the Internet Council to a board of cybercafé operators. Their appropriation of Mustafa Kemal Atatürk's image on their website to align themselves with his nation-building project exemplifies this point. The 1923 founder of the Turkish Republic, Atatürk ("Father-Turk") is the most prolific national symbol in public space. For ninety years his image has blanketed class-rooms and office walls, money, billboards, television commercials, clothing, monuments, parades, and other public signage—an icon of Kemalism, a nation-alist ideology combining militarism, laicism, and modernization.[25] For years, IIKO's "Objectives" page displayed an image of Atatürk dressed in a suit and tie pointing to the union's aims, invoking the sense that cybercafé operators were continuing his modernization efforts by extending ICTs to the masses.[26] While distinct from the AKP's religious address, IIKO's Kemalist approach resonates with the AKP's use of a "traditional," paternalistic, Muslim family to narrate its Information Society development program, e-Transformation Turkey:[27]each story frames digital technology as buttressing patriarchy, and vice versa, in the twenty-first century.

IIKO's ideological overlap with the AKP was also made clear by their 2011 launch of an alternative Internet filtering system, which they offered for free download on their website. Their logic was that if they replaced the govern-ment's current system—ironically entitled "Safe Net: Choice is Freedom"[28]— with their own filtering system, they could transform the operator's public im-age from criminal to safeguard. As Çetin explained, "We have more expertise about newly arising pornography and terror-related websites. We catch them first, before the BTK or anyone else can. Ideally, each day, when a problematic site is accessed at one of our union-member's cafés, that operator will report the information to IIKO's headquarters, who will then add it to the filter." His vision of an army of cybercafé operators monitoring cybercrime and managing the national filtering system more effectively than any corporation or govern-ment mobilizes the cybercafé's position as a critical node of ICT infrastructure to vie for greater authority. Closer to users' Internet practices, the operators are presumed the most equipped and adept at understanding and regulating them. Due to political pressure and lack of funds, however, IIKO's alternative filter was discontinued in 2012.[29]

An examination of cybercafé scapegoating and unionization reveals the different interests at stake in Turkey's PIAP regulations. Ongoing battles be-tween IIKO and the government do not signal a marked difference in opinion about how to control ICT infrastructure—both sides support centralized control. Rather, what these battles signify is a disagreement over the kinds of public val-ues that should be fostered by the ICT infrastructure: Muslim piety, global cos-mopolitanism, and economic liberalization, or secular nationalism, economic protectionism, and state-centric modernization? In Turkey the government and the media largely shape how cybercafés are perceived in the public sphere. Yet by focusing on how cybercafés are politically disruptive and morally corruptive, these publicized perspectives overlook how cybercafé operators' labor practices are fundamental enablers of ICT infrastructure for the information have-less. In the next section, I will address how low-income and young people in Turkey understand ICTs as tools for political dissent and socioeconomic empowerment via the cybercafé operator's practices of repair, circumvention, and ambivalence.

Circumvention at Cybercafés

Turkey's anxiety-ridden media imagery of moral, spiritual, and national demise at cybercafés reveals these sites as vulnerable nodes of ICT infrastructure that disobey Internet content and copyright policies. Despite their infamy, cybercafés have proliferated in Turkey—reaching approximately twenty-five thousand locations by 2012—because they meet working-class users' demands for greater access. In nonunionized cybercafés, operators regularly bend the rules in order to create an ICT infrastructure for the information have-less. My fieldwork revealed banal circumvention—from software piracy to proxy use to quickly discarded surveillance logs—in cybercafés across the country, a stark contrast to the unions' spotless PR depictions. In this section, I analyze the cybercafé as an amorphous space of instability and stability. By regularly failing or refusing to purchase software licenses and implement filtering and surveillance measures that drain budgets and bandwidth, operators impede government and corporate interests. At the same time, the operators coordinate, repair, and teach in order to fill an infrastructural void for have-less users. Invocative of Stephen Graham and Nigel Thrift's assertion that repair "continuously surrounds infrastructural connection, movement and flow" and "can itself be a vital source of variation, improvisation and innovation,"[30] the generativity emerging from operators' circumventions are a prerequisite for working-class network society. While I regularly observed operators repairing equipment malfunctions, here I will conceptualize another form of repair that is prevalent in cybercafés—operator-facilitated circumvention of filters and copyright. Hacking and piracy are forms of infrastructural repair that we must take into account when studying PIAPs in economically underprivileged contexts.

In Turkey the cybercafé is a space for all kinds of repair—from tinkering, recombination, and troubleshooting, to piracy and proxy use. Although disruptive for regulators and corporations, these practices are generative for the information have-less. Particularly in cybercafés located in urban peripheries, smaller cities, towns, and villages, and in poorer urban districts, the rules delineating age limits, hours of operation, smoking, security cameras, surveillance, and website filtering were regularly bent by operators. Meanwhile, gaming, social networking (Facebook, Twitter, MSN chat), p2p file sharing, and banned sites (via proxies) filled customers' screens—not school research and homework, as IIKO purported. To mitigate blocks or lags in connectivity and to access the desired content, operators and customers—who in smaller towns have social ties outside the cafés—collaboratively build their technological skills and knowledge through trial and error. When facing website blocks and application incompatibilities, they help each other, innovate solutions, or find fixes through online forums. Circumvention repairs/generates infrastructure for working-class users and also cultivates a form of technological literacy that highlights the plasticity of infrastructure (filters can be bypassed, blocked content can be accessed). In contrast to the discourse of scapegoating, from the vantage point of the have-less, cybercafé operators are essential facilitators who either directly help or passively allow users to navigate around access barriers, decipher unfamiliar interfaces, languages and codes, participate in an Information Society that excludes them, and create alternative information networks of their own—such as online communities that organize around GLBTQ, ethnic minority, and women's issues.

Is piracy necessary to build a have-less ICT infrastructure when open-source options are available? Most operators agreed that open source was a viable option in theory; however, they pointed out that in practice it was not beneficial for their customers, who can secure better jobs if they are adept at proprietary software. "What good would it do for our children to learn Pardus when the world is using Windows?" an operator in Istanbul asked me, comparing Windows to Pardus, an open-source operating system developed by Turkey's national science foundation (TÜBİTAK).[31] "It's incompatible with what users in the U.S. and Europe have."[32] His collapse of "world" with the United States and Europe was part of a shared tendency among operators to use trends in Western technological practices as a development measuring stick for Turkey. On their storefronts, operators frequently advertise Windows to attract students who use the same software at school or through the national computer-tablet FATIH Projesi (FATIH Project), which aims to distribute millions of tablets equipped with Windows 8 to Turkish schoolchildren from 2012 to 2017.[33] In contrast to Pardus, Ubuntu (Linux), and other open-source options advocated by elite service providers (including college-educated hackerspace organizers whom I interviewed in 2012), access to Windows supports a dream shared by most families in Turkey: to enable their children to pass through the educational bottleneck and gain the technological skills needed to become employable. Despite their more limited socioeconomic mobility, the have-less aspire to achieve professional success by learning how to integrate and use Western operating systems at cybercafés.

Because their customers demand proprietary software, operators often resort to piracy in order to afford access. In Turkey, cybercafés, small DVD and electronics shops, and mobile-phone stores collectively create a network for unlicensed software and hardware exchange. Cybercafé operators fit into this network by providing pirated software on their computers and by installing pirated software on their customers' PCs. The latter service extends their reach to a middle-class customer segment who would not come to cybercafés to access the Internet or a computer but do come to install the latest version of Windows on their laptops for an affordable price. Figure 9.3 shows an operator in his late teens installing pirated software on a middle-aged customer's laptop. Having had difficulties with her current OS, this customer came to the café after hearing that someone there would install Windows 7 for a lower price than market cost. During the transaction, there was no mention of piracy nor the impression that the operator or customer thought their exchange was criminal. The loyalty emerging from entangled social and kinship networks between operators and customers in small towns, coupled with high demand for these products, facilitates a robust underground market that is extremely challenging to regulate. Because piracy at cybercafés is dispersed and difficult to identify, it remains ignored by corporations and police. As Microsoft Turkey's piracy prevention director explained it, there is neither the will nor the resources—within the government or at Microsoft—to enforce intellectual property law in all twenty-five thousand of Turkey's cybercafés.[34] This is why police enforcement oscillates from fines and café closures to ID checks and verbal warnings. In fact, Microsoft currently has financial incentive to temporarily *allow* piracy in order to saturate Turkey's emerging have-less market with Windows OS and stave off competition.

In addition to avoiding the term "piracy," cybercafé operators also avoid the term "hacking," which is associated with cyberterrorism by the government and the media. To be clear, most operators are hackers. They assist their customers in "hacks," including rerouting around website blocks, grappling with slow speeds, navigating unfamiliar applications, reformatting files, searching for sensitive information online, utilizing open-source solutions, publishing blogs, and locating/downloading proprietary applications. I am not including images of operators here, and I have changed the names of all of my interview subjects, but what would be clear from any images that I did share is that most operators are under age thirty. They grew up in an era when cybercafés, gaming cafés, and the Internet were accessible to boys and men of diverse economic backgrounds (late 1990s onward). They have tinkered with computers for most of their lives and typically start their careers as apprentices for older operators, but after some time they start enterprises of their own. Operators in Turkey understand youth culture because they are living it (or recently lived it) and are thus sympathetic to their young clientele who demand greater online freedom. This encourages an environment where—if not directly engaging in circumvention—operators overlook customers' piracy and proxy use despite the laws mandating surveillance and filtering and despite conservatives' disapproval. Operators' user-logging monitors (figure 9.2) primarily serve a performative function—to reassure concerned families and prevent police raids. When I asked for how long surveillance logs were kept, operators' responses ranged from twenty-four hours to one year (the latter being the legal requirement when I conducted the interviews). Even when operators do apply the legally required filters and surveillance measures, they often do so in a perfunctory fashion because they understand that customers will find new proxies to bypass these constraints, which will in turn require them to spend more of their limited resources (personnel, servers, software) on updates.

From the perspective of the operators and their young customers, cybercafés are spaces of repair and entrepreneurism via circumvention (hacking and piracy). Bypassing website filters and accessing proprietary goods, customers are either directly helped by the operators or indirectly encouraged by their ambivalence. Despite official prohibitions, the operators enable circumvention in order to sustain and grow their businesses. Put another way, circumvention is

Figure 9.3. A cybercafé operator installs unlicensed Windows 7 on an older patron's laptop. The operator's finger is dyed from a henna ceremony at a recent wedding. Turkey, 2011. Photo by author.

the process by which the cybercafé functions by malfunctioning: the operators' disruptions of the law generate and extend an ICT infrastructure to working-class users for whom it otherwise would not exist. Criminalized practices of hacking and piracy are an integral part of how cybercafé customers understand, access, and communicate through the ICT infrastructure.

Navigating Social Protocol

As the previous sections describe, cybercafé operators are stuck between regulations they cannot afford to follow, a national media that scrutinizes them, and a burgeoning demand by the information have-less for access. In this context, ICT infrastructure is comprised not only by material technologies but also by circumvention practices that enable underprivileged groups to participate in the Information Society. In this final section, I explore how ICT infrastructure and PIAP access are also influenced by gender and religious differences. In pious and conservative areas in Turkey, there is a tension between the sociocultural protocol restricting women's access to cybercafés, and cybercafé operators' interests in profiting from this untapped female market. Despite Turkey's relatively stable economy and global reputation as a "successful Muslim democracy," gendered asymmetries in educational and economic opportunity remain. In the World Economic Forum's Global Gender Gap Report (2012), Turkey ranked 129 out of 135 in the category of women's "Economic Participation and Opportunity," placing them among Iran, Yemen, and Saudi Arabia, a surprising figure for this E.U.-candidate member and one of the first countries in the world to legislate female suffrage.[35]

As Mutlu Binark has theorized, the cybercafé extends the tradition of the *kahvehane*, a Turkish coffeehouse where men gather, drink tea and coffee, smoke cigarettes, read newspapers, play backgammon, socialize, and discuss politics.[36] These public spaces catering to male socializing are commonplace; in cities large and small, male kahvehane are found every few blocks. Although there are a handful of female and coed kahvehane in Istanbul, Izmir, and Ankara, both kahvehane and cybercafés that serve women are scarce, and particularly so in smaller cities and towns. Like the kahvehane, the cybercafé is generally regarded as a male space, and as discussed earlier, the press depicts cybercafés as bastions of lewd material. In our interviews, all operators agreed that relentless press coverage linking cybercafés to violence and pornography had fueled these impressions; an infamous example occurred in 2007, when Hrant Dink, a nationally beloved journalist and the controversial editor-in-chief of the Armenian newspaper *Agos News*, was assassinated by a teenager who was rumored to have planned the murder in a cybercafé.[37]

Particularly in cybercafés located in central and eastern Anatolia, but not limited to this region, social protocols prescribing family duty, national loyalty, neighborliness, honor, and gender-appropriate behavior are just as critical to the extension of ICT infrastructure as technological protocols that determine packet routing, licenses that determine software/hardware compatibility, and server/bandwidth capacities that affect the speed of transfer. In addition to their business and technological acumen, cybercafé operators must persuade pious families—long familiar with scapegoating news coverage—to allow their female relatives to become customers. Most of the operators I interviewed were raised in the towns where they worked and were familiar with cultural sensitivities pertinent to their communities. Invoking Downey's description of telegraph messengers as "mediating between the customer and the rest of the telegraph network," operators must "speak intelligibly" in order to assuage local concerns and negotiate a space for ICT engagement.[38] In addition to mediating between the infrastructure and customers through circumventive repair, operators are cultural translators, interfacing between their communities and technologies depicted as threats to tradition, religion, and national stability.[39]

An example of cybercafé operators' cultural mediation emerges through their interior design and scheduling choices that encourage female participation. As I described in the introductory anecdote, some cafés are partitioned into male and female sections, providing a separate, private space for female customers. More commonly, the operators who do not have the resources for a complete partition will seat female patrons at computers located near the front and offer computer-station partitions to provide each customer more privacy. Operators also promote female-only or female-friendly schedules to encourage female participation during daytime hours. One operator I interviewed painted his café pink and purple in order to "send the message that women are welcome."[40] Seeking to address and foster the female market, these operators explained how their main obstacles are local attitudes framing virtuous women as dutiful, home-bound wives, mothers, and daughters. Accordingly, a "good" woman honors her husband by fulfilling household duties and caring for the children; a "good" daughter assists with homemaking and lives at home until married.

Reality contrasts with this attitude, as many women in Turkey work outside the home. The economic liberalization and urbanization of labor in the 1980s fostered unprecedented female movement to cities for education and employment opportunities so that they could independently earn income and send money home. Also, women in Turkey do move around in public spaces—in eastern and western regions, in small towns and big cities, as students, workers, leaders, and community members. Jenny White has researched how the headscarf, or *teşettür*, offers working-class, pious women greater physical mobility in public space, expanding their political and economic agency.[41] At the same time, there is an undeniable reality that in Turkey, particularly in conservative neighborhoods or towns, if women are seen loitering in public spaces that are frequented by men, they risk injury to their honor and reputation. From the operators' perspectives, a pious sociocultural protocol prompts protective families to prohibit their daughters from coming to cybercafés populated by men, and the way to circumvent this barrier is through the separate-but-equal treatment of female and male customers.

After conducting interviews with the operators who took this approach, I became their customer, entering female-only sections or using the female-designated computers near the storefronts. In the female-only sections, I encountered lively atmospheres reminiscent of the female-friendly spaces I experienced in countless homes, in beauty parlors, and in mosques' female-only sections. Women and girls chatted with each other, sharing information and technical help. They talked to relatives and friends living abroad through online video-chat. They researched educational opportunities and job openings. They read the news. The rare cybercafé that offers privacy for women facilitates the circumvention of sociocultural protocol demarcating female mobility in public space. Female customers access a rare privacy away from their families to converse with friends and partners; they access ideas that challenge some religious beliefs (such as online content explaining evolutionary theory), and they connect to a wide variety of social networks with diverse groups (for example, Turkish GLTBQ communities are extremely active online and use the Internet as a communication and mobilization tool). Repairing their damaged reputations through innovative spatial and temporal designs, operators can distinguish the cybercafé from the kahvehane to encourage and profit from women's engagement in the ICT infrastructure.

Working to mediate between working-class, pious, and female customers and an ICT infrastructure more accessible to the wealthy and educated, cybercafé operators have the power to affect technological literacy in profound ways. They can shape how a vast new user group—have-less women—understands and connects to ICT infrastructure. The cybercafé unions define technological literacy as Turkish citizens' ability to navigate Western proprietary software and access culturally and nationally appropriate online content. Nonunionized cybercafé operators would add to this definition the ability to utilize circumvention tools, to pirate, and to recycle, reuse, and improvise when ICTs are in disrepair. Finally, while both unionized and non-unionized operators prioritize the extension of technological literacy to male working-class users, a handful of operators—by changing their café's designs and negotiating with their communities—assert that these literacies should also be extended to women.

In this chapter I focused on a have-less technological literacy emergent within cybercafés, yet other service providers in Turkey build different meanings of technological literacy to serve the "haves." For example, while collaborating with hackerspaces in Istanbul in 2012,[42] I noted how the organizers defined technological literacy as knowledge of open-source programming and encryption. Whereas cybercafé operators teach Windows navigation, hackerspace organizers teach Ubuntu. Whereas circumvention intermittently pushes cybercafé customers toward considering protocol and code, the latter are central to "hack-a-thons," collaborative brainstorming events where hackerspace participants reengineer hardware and software. Cybercafé operators cited socioeconomic reasons for why the most meaningful form of technological literacy was front end, while hackerspacers cited political reasons—state censorship and surveillance—as to why technological literacy should emphasize back end. Nevertheless, while serving different demographics (dissident elites and working-class youth), both hackerspaces and cybercafés, whether intentional or not, effectively exclude women. Through my fieldwork it became clear that women and girls without home Internet access—namely, economically underprivileged women living in pious communities—have not yet reached the category of "information have-less." Further research can help us to identify other service providers, if any, who extend ICT infrastructure and infrastructural legibility to this group.

Conclusion

In Turkey, the cybercafé operator's practices of repair, circumvention, and extension are an essential component of the ICT infrastructure for have-less users. While Internet regulation debates and policymaking conferences ensue between governments, NGOs, and corporations, local service providers who foster and maintain public Internet access points generate, connect, and teach ICT infrastructure to working-class users who are excluded from stakeholder discussions. Ethnographic research at Turkey's cybercafés reveals how operators function as an infrastructural bridge, smoothing technological incompatibilities and access gaps through practices of circumventive repair and a refusal to enforce prohibitions. By mediating between conduits and users, navigating technological and social protocols, and shaping customers' technological literacies, cybercafé operators influence public expectations as to the roles and purposes of infrastructure. In their capacity to dial up or down the visibility of infrastructural elements—the hardware, software, networks, protocols—service providers offer their customers tools to identify and critique website blocks, surveillance, and information society exclusions. They foster regulatory malfunctions that can disrupt monopolies on political speech and in high-tech industries. They can bypass sociocultural protocols that limit access for women and licenses that limit access for the poor. As such, when studying the myriad histories, contexts, and possible futures of ICT infrastructures, it is necessary to analyze the mediating functions of local service providers alongside the louder and more visible stakeholders. Making service providers visible requires ethnography as an integral part of the research methodology. Identifying and analyzing intermediary workers at the endpoints of infrastructure cannot be gleaned from network diagnostic tools that measure conduit speed and functionality, nor will service-provider labor become legible through legal or political economic accounts. Combining these methods with multi-sited ethnography—extensive dialogue with service providers and participants—will help us build a more complete picture of how technological conduits, human practices, and natural environments interact to generate ICT infrastructure.

Notes

My deepest gratitude to Lisa Parks, Vijay Kumar S., Nicole Starosielski, and Cristina Venegas for their intellectual generosity and feedback on earlier versions of this chapter.

1. For exact numbers, see "Percentage of Individuals Using the Internet" (Geneva: International Telecommunications Union: 2012); and "Istatistiklerle Türkiye 2012," pub. no. 3942 (Ankara: Turkish Statistical Institute: 2012).

2. Cities included Istanbul, Gaziantep, Sanliurfa, Diyarbakir, Erzurum, Ankara, Trabzon, Samsun, Izmir, and Eskisehir.

3. "Steps toward an Ecology of Infrastructure: Design and Access for Large Information Spaces," *Information Systems Research* 7, no. 1 (1996): 111–34, 113.

4. See Miyase Christensen, *Connecting Europe: Politics of Information Society in the EU and Turkey* (Istanbul: Bilgi University Press 2010).

5. Jack Qiu, *Working-Class Network Society: Communication Technology and the Information Have-less in Urban China* (Cambridge, Mass.: MIT Press, 2009), 9.

6. This is on the low end of estimates, taken from Turkish Statistical Institute data in 2013: http://tuik.gov.tr/PreHaberBultenleri.do?id=13569 (accessed February 10, 2014). Other data show up to 75 percent of Internet users in Turkey regularly accessing the Internet from PIAPs. See Ahmet Eskicumali, "The Effects of Internet Cafes on Social Change in Turkey: The Case of Hendek," *Turkish Online Journal of Educational Technology* 9, no. 2 (April 2010), available at http://www.tojet.net/articles/v9i2/9220.pdf (accessed February 10, 2014).

7. Gregory J. Downey, "Telegraph Messenger Boys: Crossing the Borders between History of Technology and Human Geography," *Professional Geographer* 55, no. 2 (2003): 134–45, 141.

8. Data cited from *Engelli-Web* ("Blocked-Web"), an anonymously run database collecting statistics on Internet censorship in Turkey. See http://engelliweb.com (accessed February 10, 2014).

9. Representative headlines include "AKP Deputy Learns about a High School–Age Homicide Suspect's Internet Cafe Habits," *Radikal News*, March 26, 2006; "Internet Café Porn Operation," *CNN Turk*, November 28, 2008; "Municipal Police Inspect Internet Cafes in Order to Protect Children," *Zaman*, October 1, 2008.

10. "5651: On Regulating Broadcasting in the Internet and Fighting against Crimes Committed through Internet Broadcasting," Turkish Prime Ministry, May 2007, available at www.tbmm.gov.tr/kanunlar/k5651.html (accessed March 30, 2013).

11. The BTK (*Bilgi Teknolojileri ve Iletisim Kurumu* or "Information and Telecommunication Authority") is part of the Information and Communications Technologies Authority (*Telekomünikasyon İletişim Başkanlığı*, or TIB), an executive-branch authority regulating telecoms in Turkey. It was formerly known as the Telecommunications Authority, or TK, *Telekomünikasyon Kurumu*.

12. The YouTube ban from 2007 to 2010 was initiated after a Web video called the founding father of the Turkish Republic, Atatürk, a homosexual. According to the TIB, YouTube remained blocked because Google refused to remove the video.

13. "Turkey," *Open Net Initiative*, December 18, 2010, available at http://opennet.net/research/profiles/turkey (accessed February 10, 2014).

14. See T. Arango and C. Yeginsu, "Amid Flow of Leaks: Turkey Moves to Crimp Internet," *New York Times*, February 6, 2014; Elif Akgul, "How Will Turkey Be Censored on the Net?" *Bianet*, January 31, 2014.

15. *Internet Toplu Kullanim Saglayicilari Hakkinda Yonetmenlik No. 26687* (Regulation on the Providers of Public Use Internet, Law No. 26687), Turkish Prime Ministry, November 1, 2007, available at http://www.siyamiozkan.org/mevzuat/27649.html (accessed September 20, 2014).

16. Two similar cybercafé unions have formed—one in Ankara, the capital city, and the other in Diyarbakir, a southeastern city known for its sizeable Kurdish population.

17. Personal interviews with Emre and Çetin (changed names), November 10, 2011, Istanbul, and December 2, 2011, via email. Author's translation.

18. Turkish households with Internet access at home reached 47.2 percent in 2012; households with computers at home, 50 percent; individuals using the Internet, 45.1 percent; mobile-broadband subscriptions per one hundred inhabitants, 16.3. "Turkey Profile," International Telecommunications Union, 2012, available at http://www.itu.int/net4/itu-d/icteye/CountryProfile.aspx#Europe (accessed September 20, 2014).

19. A Microsoft Turkey representative noted that first-time piracy offenders are typically fined at three times the market value of the products they pirated.

20. "Bilgi Toplumu Istatistikler, 2010" ("Information Society Statistics, 2010"), T. C. Basbakanlik Devlet Planlama Teskilati Mustesarligi, June 2010, available at www.bilgitoplumu.gov.tr/Documents/1/Yayinlar/BilgiToplumuIstatistikleri_2010.pdf (accessed September 20, 2014).

21. *Sayili Fikir ve Sanat Eserleri Kanunu 5846* (Law on Intellectual and Artistic Works), Turkish Prime Ministry, December 5, 1951 (last amended in 2008), available at http://www.wipo.int/wipolex/en/details.jsp?id=10675 (accessed January 21, 2013).

22. Turkey has signed TRIPS—the Agreement on Trade Related Aspects of Intellectual Property Rights—a global IP agreement started in 1994 and administered by the World Trade Organization holding national governments accountable for enforcing international copyright: see http://www.wto.org/english/docs_e/legal_e/legal_e .htm#TRIPs.

23. Personal interview with Microsoft Turkey's Director of Piracy Prevention, March 14, 2013.

24. IIKO president addresses the Internet-freedom activists in a blog post after their mass public demonstrations in 2011: "What are your intentions? Have you sought out your own solution or researched this further? No, you have not sought answers, and without researching, without information, without learning, you are shouting and marching! In the future, more government prohibitions are coming. Friends, who has ever tear-gassed you? Have you ever searched Google for stories about sex? Have ever you seen child pornography advertisements pop up on opened websites? Do you know what your children are able to learn from these pornographic stories and images? *Which "freedom" are you talking about?* We, as Internet café operators, condemn any policymaking that risks our children's mental health and futures." Available at www .iiko.org (author's translation).

25. Sibel Bozdoğan and Resat Kasaba, eds., *Rethinking Modernity and National Identity in Turkey* (Seattle: University of Washington Press, 1997).

26. See http://www.iiko.org.tr/kurumsal-kimlik/ilkelerimiz (2009–2012). The image I discussed was removed in 2011 but has been archived at www.proxyculture .wordpress.com.

27. "E-Transformation Turkey: Information Society Strategy and Action Plan (2006–2010)," Turkish Republic Prime Ministry, State Planning Organization, July 2006, SPO Publication No: 2700, available at http://www.bilgitoplumu.gov.tr/ Documents/5/Documents/Action_Plan.pdf (accessed April 10, 2013).

28. *GüvenliNet: Seçmek Ozgurluktur* ("Safe Net: Choice is Freedom"), Turkish Information and Communication Technology Authority, available at http://www.guvenlinet .org/tr.

29. An explanation of the financial and regulatory pressures causing IIKO's termination of their filtering initiative can be found in Turkish at http://www.iiko.org.tr/ ik/110-oda-projeleri/proje2/123-guvenli-webfiltre-projesi-durduruldu.

30. Stephen Graham and Nigel Thrift, "Out of Order: Understanding Repair and Maintenance." *Theory, Culture and Society* 24, no. 3 (2007): 1–25, 6, 17.

31. Pardus is an open-source OS developed for Turkish users by the National Science and Technology Foundation of Turkey (TÜBİTAK), available at http://www .pardus.org.tr.

32. Personal interview with anonymous cybercafé operator in Istanbul, November 20, 2011.

33. *FATIH Projesi: The Movement to Increase Opportunities and Develop Technology Project*, Turkish National Education Ministry, 2011, available at http://www.fatihprojesi.com.

34. Personal interview, Director of Piracy Prevention, Microsoft Turkey, March 14, 2013.

35. Turkey ranked 108th in education and 129th in economic participation in the World Economic Forum's *2012 Global Gender Gap Report*. See http://www3.weforum.org/docs/GGGR12/MainChapter_GGGR12.pdf; in terms of the country's reputation in Western press, see, for example, "Turkish Democracy: A Model for Other Countries," on *All Things Considered*, National Public Radio, April 14, 2011, available at http://www.npr.org/2011/04/14/135407687/turkish-democracy-a-model-for-other-countries.

36. Mutlu Binark, "How Turkish Young People Utilize Internet Cafés: Results of Ethnographic Research in Ankara." *Observatorio Journal* 8 (2009): 286–310.

37. Each year in January, Dink's 2007 assassination is nationally remembered. It is believed that state officials were involved in Dink's murder and have not been held accountable: see http://bianet.org/english/minorities/131762-hrant-dink-murder-trial-summary-of-discrepancies (accessed September 20, 2014).

38. Downey, "Telegraph Messenger Boys," 140.

39. Prime Minister Tayyıp Erdoğan's June 2013 declaration that Twitter is a "menace" to society is one example among his many warnings of digital technology's threat to traditional culture and national unity.

40. The purple cybercafé is located in downtown Gaziantep—a city in southeastern Turkey along the Syrian border—owned by a pious male operator in his early thirties.

41. Jenny White, *Islamist Mobilization in Turkey: A Study in Vernacular Politics* (Seattle: University of Washington Press and the Institute of Turkish Studies, 2002).

42. In 2012, at Istanbul Hackerspace, one of two hackerspaces started in Istanbul that year, anywhere from ten to thirty people attended programming classes, and approximately twenty to seventy-five attended outreach events (guest lectures by computer science professors and industry participants, digital kite-flying, robot building). See https://istanbulhs.org.

CHAPTER 10

The Internet as the Anti-Television

Distribution Infrastructure as Culture and Power

CHRISTIAN SANDVIG

In a prank circulating on the Internet in 2015, the victim is presented with a link to a video with an attention-grabbing title. When clicked, the screen shows the familiar rotating circular pattern of dots that convey the video is loading: a "wait indicator" in the jargon of human-computer interaction. The text "Buffering Video . . ." also appears. This video is actually a looped shot of the wait indicator itself. There is nothing but the wait indicator. In one YouTube version of this prank, a commenter wrote: "This must be the most watched thing on all of YouTube."[1]

If the reader feels the pain of this commenter, he or she might be surprised to know that videos buffer for reasons that are quite different from those most viewers expect. This chapter investigates the invisible infrastructure that delivers video over the Internet and argues that the availability and quality of video on the Internet are significant new political and economic battlegrounds where culture is controlled. The case of Internet video distribution also makes clear that the *infrastructure of distribution* is a crucial site for the analysis of media technologies. Focusing on infrastructure (after Star[2]) is also an essential task for those who hope to know and to change media and technology.[3]

Distribution asks us to revisit a classic question of media studies: How does the medium affect the content? This chapter will demonstrate that the Internet was originally conceived of as the opposite of television: the anti-television. Over the course of several decades, however, the Internet was technologically retrofitted and transformed to make video distribution possible. Embedded in this transformation were competing ideas about what content and which audiences are valuable, and indeed how culture itself ought to work. The selection of videos available on the Internet today—and how that video looks—result from purposeful decisions made by actors who hoped that either the model of television or the Internet as anti-television would prevail.

The Beginning: Point-to-Point

Technically speaking, television and Internet traffic were at first like oil and water: fundamentally unmixable. The Internet was envisioned as a "point-to-point" network,[4] meaning a system designed to facilitate communication between two nodes. Although some functionality in the Internet protocols allows the broadcast of data to all nearby nodes,[5] uses like broadcasting content to a large audience were never envisioned by the engineers who built the system. At the time, computers were not capable of receiving or displaying video at all.

In communication network design, the distinction between point-to-point and broadcast systems is one of the most basic. The metaphor used to explain the point-to-point Internet given by engineer and Internet pioneer Vint Cerf is that of the postal network, with packets of data functioning like postcards.[6] This is an apt metaphor that highlights the difference between broadcast and point-to-point. In over-the-air television broadcasting, a fundamental feature of the electromagnetic wave that radiates from a television station's transmission tower is that it makes no difference to the wireless signal whether it is received by one person or one hundred. Indeed, the cost of sending it—the cost of transmission—is the same in either case. Delivering television via satellite or via a cable network also employs a broadcast architecture and realizes the same benefit. In contrast, the costs of transmission for a postcard (and the Internet) scale linearly: in the case of one versus one hundred recipients, ninety-nine more postcards must be printed by the sender, and ninety-nine more postcards must be delivered by the mail carrier. Mail carriers must do ninety-nine times more work, but television antennas need change nothing. Ninety-nine more postcards require ninety-nine more stamps. In other words, unlike television broadcasting, the cost of transmission rises as the number of postcards transmitted increases.

The Internet could certainly have been designed differently,[7] but a postcard-like system conformed with the design goals of early Internet engineers. When I click on a link or type in a Web address to read a news story published by the *New York Times* at nytimes.com, a reasonable person might assume that the information comes from New York City, from a computer owned by the *New York Times*. The original vision of the Internet's design presumed as much. Information that some user wanted would be found where it had been produced, and the network's job was to facilitate a connection between one source and one recipient. This is partly because the point-to-point system linked relatively expensive, powerful, multipurpose computers that could act equally as senders and receivers—unlike the "dumb" televisions of the time that could only receive. All devices on the early ARPANET, the precursor network to the Internet, were expensive, multipurpose computers. In 2013 an Internet router cost about thirty dollars and was at least ten times cheaper than a computer. But on the ARPANET what is now called a router was a full-fledged computer in its own right. The first router, a Honeywell 516 microcomputer, was six feet tall and had four steel eyebolts in the top so that it could be transported by helicopter; it cost $100,000 in 1969[8]—$634,000 in 2013 dollars.

The users of this pre-Internet system (the ARPANET) were homogenous: they were largely computer scientists at elite educational institutions, and there were not many of them. At its launch in 1969 there were just four nodes on the network. Fifteen years later the network reached one thousand nodes. The foundations of the protocols we know as the Internet were crafted to serve a network of a few hundred computer experts using very capable, expensive machines. The largest, most successful, and largely unforeseen use for this system was a point-to-point application: email. (A 1973 report estimated 75 percent of the ARPANET's use was email.[9]) The Internet's designers envisioned using these machines to facilitate research file transfers, or, later, text-only email.

A postcard-like system for handling these communications was probably the most logical choice because it presented interesting technical problems in the context of computer networking research in the 1960s, and the ARPANET was a research network.[10] When it became clear that the ARPANET would become a network for non-academics, the envisioned users did not seem like people who would want "mass" communication. Personalization was assumed; users would all want different information. The early Internet was to be a network of equals, with the ideal user thought to be producing new knowledge, not passively receiving it. "Laudatory descriptions of the word 'active' in discussions of media use" have a long, gendered, and problematic history.[11] The Internet's pioneers were enthralled by what Nathan Ensmenger calls the myth of the "super-programmer"—a white-collar, well-paid male computer professional envisioned as an elite knowledge worker.[12] The users of the early Internet were thought to be "autonomous and creative,"[13] and the future network was depicted as serving elite men in universities and in industrial settings like IBM. In this future network, users would also be producers, content would be plentiful, and attention to it would be widely distributed.

Lick's Television: The Opposite of Television

When these ideas about the Internet were conceived, television could not have been more opposite. In the 1960s, television was a broadcast medium designed to distribute a show like *Gidget* or *Gilligan's Island* from one source (a television network like ABC or CBS) to as many receivers as possible. Television's one-way distribution network consisted of relatively unsophisticated nodes (television sets). In the 1960s the average U.S. television household received fewer than five channels, "cable" television referred to a cable that extended the range of an antenna, color television was new and not widespread, and "premium channels" did not exist. In the past, industry commentators often framed the rise of the Internet as a challenge to the network architecture and ideas of traditional telephone companies as the "netheads" versus "bellheads."[14] The canonical history of the Internet[15] does not mention the word "television" a single time. Despite that, it is this contrast of "television versus Internet" distribution that has come to define the media industries today, and it is this conflict that will ultimately come to transform both combatants.

What is at stake is not some arcane technical principle of point-to-point routing versus broadcasting, but the shape of culture itself. The Internet was the anti-television, and one of the pioneers of the Internet said as much in 1967. J. C. R. Licklider, a psychologist who headed the Information Processing Techniques Office at the Pentagon, is now credited with promoting a vision of computing that would become the Internet. Licklider, often known as "Lick," convinced the U.S. government to fund such a system and created interest among the engineers who would invent the means to make it possible. By one account, "most of the significant advances in computer technology . . . were simply extrapolations of Lick's vision . . . he was really the father of it all."[16]

Licklider's influential paper, "Man-Computer Symbiosis" (1960),[17] specified how interactive computing ought to work long before it was technically possible. Less attention has been given to his other visionary writing. In the late 1960s Licklider was invited to prepare a research paper for the Carnegie Commission on Educational Television, an influential nonprofit research and policy body whose proposals eventually led to a significant reorganization of television broadcasting in the United States and to the founding of the Corporation for Public Broadcasting. Licklider's research paper, "Televistas," did not receive wide attention.[18] In it he issued a stinging indictment of the existing technological system of television. He based his critique almost entirely on the system's distribution and transmission characteristics, writing,

> The great simplifying characteristics of conventional broadcast television are that it is broadcast and that the broadcast stations transmit to viewers who do not transmit back. . . . From an educator's point of view, the main intrinsic defects of broadcast television are that it offers everyone the same thing and does not give viewers a direct way of participating . . . [19]

Licklider went on to assert that what he called "selective television," which involved interactivity, would soon be possible via computer networks. In a dizzying feat of prediction, he forecast the end of "liveness" as a distinguishing feature of television and suggested that everyone would be able to select their own programs in near-real-time, watching them almost instantly. He foresaw a store-and-forward architecture for distributing video that is very similar to what is in use on the Internet today. He emphasized that "we are used to thinking of the output of a television set as ephemeral pictures,"[20] but that this would soon change, as television will be stored and manipulated as a data file: he called this "hard copy television."

For Lick the transmission architecture was a moral choice. His concerns were unabashedly paternalistic. He advocated for a television that would broaden access to a highly classed version of high culture, giving examples such as the symphony orchestra and community theater groups, as well as—somehow—fighting the war on poverty. He explained that this was "based on a philosophy that appreciates the interaction value of diversity among the personalities, interest patterns, of individuals as well as the cohesion value of community in language and cultural heritage—and a philosophy that prefers active participation to passive observation."[21] This was to be achieved by a global network of interconnected computers—what would become the Internet. In other words, in 1967 Licklider offered the Internet as a salvo aimed at the heart of television—its network architecture. Combat was joined, but from today's vantage point it appears to be television's distribution and transmission system that will prevail.

The Challenge of Asymmetry

Licklider's emphasis on selectivity and knowledge production promoted a future in which different users wanted different things—discussed today as the Internet's "long tail."[22] While this was an attractive story for many commentators, it often did not fit the pattern of how users actually behaved on the Internet, causing a variety of problems well before the advent of online video. When Sir Tim Berners-Lee invented the World Wide Web in 1991, like the ARPANET, it was framed as a tool for a select group of highly educated knowledge workers who would produce as much as they consumed.[23] The Web's original blueprint included the feature that any Web user could edit any Web page,[24] which now seems quite impractical. Early Web clients were referred to not as a "Web browser" but as a "browser/editor." So were all computers created equal.

Returning to the *New York Times* example, Lick's vision of the Internet and Berners-Lee's vision for the Web meant that a computer at the *New York Times* headquarters building would hold the news stories as Web pages to be disseminated, and when a user wanted one, he or she would query it (for example, from Ann Arbor, Michigan). The network's logic presumed that the same linkage might occur in reverse. The New York City computer would then potentially be used by the reporters there to query a computer in Michigan for some Web pages of value to the *Times*. In this hypothetical example, when we try to think of what kinds of Web pages an average user might write that a *Times* reporter might need, imagination fails. In fact, far more people are able to read good articles than to write them. So are media producers and audiences created unequal.

This asymmetry created serious problems when the Internet began "mass" communication—distributing the same thing to a large number of people. In a point-to-point system each communication is a separate transaction (recall, just like a postcard). A Web *server* is a machine, online all the time, that waits for a request to see one of the Web pages stored on its hard disk (so that it can *serve* them like a waiter at a restaurant, hence the name "server"). At small numbers of requests per minute, the number of people requesting a Web page from the server does not matter. But at some point, as traffic increases, the Web server or the network near the server becomes overwhelmed. Either there is not enough processing capacity to make a new copy of the requested Web page for every user who demands it (called "server load"), or there is not enough available network capacity near the server to deliver copies of those pages ("source congestion"). Remember that unlike traditional over-the-air television broadcasting, a new transaction must be made for each request.

This problem is common enough that a new phrase was coined to describe it: "the Slashdot Effect."[25] It is named after a popular 1997 technology news service on the Web called Slashdot. Slashdot invited users to submit their own links to interesting websites.[26] When a Slashdot user found a juicy Web page and shared the prize address, however, the clicks of Slashdot readers would generate requests that would overwhelm the target Web server. The act of promoting content to even Slashdot's modest audience sometimes caused that content to become instantly inaccessible due to server load or source congestion: the Slashdot Effect. Although it was named after this niche Internet service, the "Slashdot Effect" became a generic term; a large-enough massing of attention on the Internet focused on any single website would bring it down.

Television has been explained as unique in that it is a system that can be used by the establishment or "the center" of society to command public attention for a communal event.[27] It is a technology defined by the experience of millions of people all watching the same thing at the same time. But on the Internet produced by Lick's vision, such a pattern of communication was impossible. The Slashdot Effect would cause the server to crash or the network to collapse. The Internet has often been characterized as inherently amenable to decentralized communication, lateral connections, bottom-up user power, and user-generated content. Nevertheless, many commercial parties did not take limits like the Slashdot Effect as features inherent to the medium but as technical and commercial obstacles that could with effort and investment be overcome. Internet engineers asked: Since Internet audiences had demonstrated a desire to look at the same content at almost the same time, how can the Internet be redesigned to support that desire? At the same time, media companies and start-ups asked: Who will be the provider of this content that everyone wants to watch? Lick's network had challenged television with a new distribution architecture, and television rose to respond.

Retrofitting the Internet: Streaming, Multicast, and IPTV

Even before video was a major source of Internet traffic, as mainstream media sources migrated to the Web they desired large audiences and therefore asymmetric communication patterns. They sought a solution to the Slashdot Effect. At first, providers handled the problems of load and congestion by simply buying larger Web servers and more network capacity. Some mainstream media sources moved their Web servers into data centers operated by the largest and most interconnected Internet Service Providers (a practice called colocation), gaining the interconnection advantages of a central network location. Multiple identical servers were grouped together, and traffic was balanced between them, a practice called server farming. However, very popular content continued to produce "congestion events" that crippled service. For instance, during the previous decade in the United States, peak congestion events involving a high demand for video included the 9/11 attacks, the inauguration of Barack Obama, and Michael Jackson's funeral. Building a very expensive and robust network to handle rare, peak-load congestion events was not economical (this problem is common to many kinds of networked infrastructure[28]). In addition, as online multimedia shifted from audio to video, the larger file sizes of video exacerbated the problem.

Providing popular audiovisual content on the Internet had quickly come to look more like a factory enterprise from the Industrial Revolution than the postindustrial future that had been promised. Large investments in Web servers and IT staff, as well as giant, power-hungry data centers involving large capital investments, had all become a necessary part of publishing popular content on the Internet. The warehouse-sized printing machinery that pressed out each copy of a daily newspaper was being replaced by warehouse-sized computing machinery that pressed out and sent each batch of electrons—an instance of a Web page or video stream. Bits were substituted for ink and paper, yet the result was no less industrial in scale. Even when throwing money at the problem, the strategy of simply buying more and better servers did not seem to be working. The issue was more fundamental. Lick's network was built with the assumption that content was plentiful, and his network "appreciates the . . . value of diversity" in cultural products, but millions of users were demanding multiple copies of the same thing, and something would have to be done about it.[29]

A longer-term fix would be to undo Lick's vision, rewriting the basic protocols of the Internet itself. New protocol proposals aimed to make the network more amenable to one-to-many video. Sometimes termed IPTV (for Internet Protocol TeleVision), this solution was in the works but proceeded very slowly in the Internet's plodding technical standardization bodies.[30] Experimental efforts in Internet engineering also sought to build a new facility into the network, available to anyone, called "multicast."[31] Multicast (another computing term) is a hybrid architecture somewhere between point-to-point and broadcast in which the same item of content is distributed to a list of many recipients. Ideally, multicast would not result in the "postcard problem" of many duplicated requests to fulfill: if implemented as its designers hoped, nodes near each other would "subscribe" to a multicast, sharing the same "postcard" (that is, copy of the content) without generating a new request for every single recipient. This meant that the point-to-point Internet could acquire some of the characteristics of broadcasting—some transmission costs that would not increase as the number of receivers increases. However, in trials multicast techniques did not scale well with large audience sizes.

Other, more successful efforts addressed the way data flowed through the distribution network. "Streaming," in computer terms, is the display of media while they are still being received. Streaming was the norm for television— so much so that the word did not need to be coined—but it was a novelty in computing. During the 1980s and early 1990s personal computers and networks were not powerful enough to stream media—that is, it is unlikely they could receive or decode a stream of incoming data fast enough to simultaneously render it for the user. As computers and networks became more powerful, streaming became viable and pioneers like Progressive Networks (later known as RealNetworks) wrote new software and protocols like RealAudio to allow multimedia streaming. The first live event to be streamed over the Internet was the audio coverage of a baseball game between the Seattle Mariners and the New York Yankees in 1995, streamed by a RealAudio server.[32]

Streaming technology was useful because it improved the responsiveness of the Web for viewers of multimedia—no longer would they have to download an entire file before playing it. At the same time, it later offered advantages to the Web's distribution system. By determining the user's network speed, streaming software could decree that only a particular amount of data would be sent in advance of the user's need for it: this is known as the buffer. In online video distribution today, for instance, the maximum buffer size is often limited. Only a few seconds of video are sent to the user ahead of what they are currently watching. As most users watch only the first few seconds of most online videos, the rest of the video data are never sent, saving substantial network capacity.

A variety of ancillary technologies were also developed that made watching video over the Internet more tractable. Improvements in video compression resulted in new formats (such as MPEG video standards) that reduced the size of video files. Adaptive bitrate streaming, in another example, is a technique wherein a sender encodes a video at a variety of different quality levels. Poor picture quality produces smaller file sizes and thus fewer bits to transfer. In an adaptive bitrate scheme, software on the viewer's computer senses the quality of the network connection and acts as a switch directing the server to send a lower-quality version of the requested content when the network is busy, conserving network capacity. Or, to put it in the words of one user: "Netflix quality all of a sudden terrible" [sic].[33] These significant innovations in streaming and compression transformed the Internet and made it possible to reliably watch television content at all. However, the most significant change in online video distribution came with the emergence of a new kind of distribution network.

Re-Architecting the Mass Audience: Edge Caching and Upload Limits

As the Internet evolved, a remaining technical challenge was adapting its point-to-point architecture to the one-to-many asymmetries of audiences and attention. A commercial breakthrough came when an MIT applied mathematics professor created the spin-off company Akamai.[34] Rather than wait for Internet protocols to change or use custom client software (like RealNetworks), Akamai cleverly took advantage of the Internet's addressing system. The Akamai network detects where a video request originates—both in geographic and network topographic terms—and then invisibly directs that request to a server that is as close to the request as possible.[35] Unlike a public standard built into the protocols of the Internet, Akamai is a proprietary system that acts as an overlay, an invisible network concealed inside the network.[36]

This is an example of a "cache"—in computing this term means the same as it does in children's stories about pirates. A cache of pirate treasure is a place where gold has been left temporarily so that it can be picked up later. Akamai's strategy, called "edge caching," moves content away from the producers and stores it close to the consumers, reducing network load and transmission delays. This is conceptually similar to the television distribution strategy of stocking libraries of videotapes at television affiliates for local broadcast, or a local television affiliate taping a network feed, then rebroadcasting it later. For Akamai's edge caching to work, however, it would have to operate a gigantic network of data centers all over the world, putting its own servers as close to valuable audiences as possible.

Although the company has zero name recognition among Internet users, in a little more than a decade Akamai was running the largest number of Web servers of any entity in the world, with servers in eighty-seven countries, connecting nineteen hundred distinct Internet subnetworks.[37] While companies like Microsoft, Facebook, and Google probably operate more servers—their total numbers are not known—the computers at those companies also do more than act as Web servers or as a distribution system for others' content. Yahoo! was Akamai's first major customer, and other customers that followed have included Apple, Google, Disney, ESPN, and Viacom. Up to 30 percent of all Internet traffic ran across Akamai's distribution network in 2013,[38] serving more than 50 percent of the Internet's top one thousand websites by traffic volume.[39] Those large media and Internet companies that do not use Akamai likely have gone into the distribution business themselves to reduce costs, building their own network of edge caches around their most valuable audiences. Akamai's edge cache overlay technique pioneered a market that would later come to be called "content delivery networks" or sometimes "content distribution networks" (CDNs). The top three CDNs in 2013 market share were, in order, Akamai, Amazon, and Edgecast (the latter owned by telecom giant Verizon).

The operation of these hidden (to users) edge-caching distribution networks can produce surprising consequences. If a video source pays for CDN distribution, Web pages and videos will load faster and may play at a higher quality. CDN-hosted videos are less likely to be interrupted, and they are less likely to change resolution while playing. Some CDNs also offered tiered service, allowing their clients to pay more for better service. To the viewer who is not aware of the distribution infrastructure, the experience of "flow"[40] when viewing online video is quite puzzling. Discussion boards are filled with varieties of the same question: "Why do ads always load flawlessly, while other video is choppy and slow-loading?" (The answer could be a CDN.) Or, "Why do some videos look terrible on a fast Internet connection?" (No CDN.) Internet audiences have no way to know why the quality of some videos is worse than others. They are likely to wrongly blame their Internet service provider rather than to realize that their attention is less valuable than someone else's and that a producer declined to pay to make this video load faster for them.

The marketing literature for CDNs claims that a video producer subscribing to a CDN will see a 60 percent to 99 percent reduction in the network bandwidth they use (users now query the CDN, not the source). CDNs promise responsiveness that is seven times faster (or greater) than content from nonsubscribers. CDNs are also facilitating a new kind of performance-based differentiation in Web content. Even though Web pages themselves continuously become larger and more complex, CDNs now measure average response times in milliseconds and are aware that the online audience can be trained to differentiate these load times and to desire a particular user experience: they can be trained to notice and appreciate CDNs without knowing that they exist. These are the kinds of production values that have long been used by well-financed players for competitive advantage in the media industries.[41]

In a more worrying vein, until Amazon entered the CDN market with its CloudFront offering in 2008, the best CDNs (including Akamai) refused customers not affiliated with major corporate content producers.[42] Although this echoes the "corporate liberalism" of earlier U.S. broadcast policy, which restricted the television medium to major producers,[43] in this case the motive was probably that of a wholesaler (the large CDNs) uninterested in the retail trade. Until Amazon's entry, smaller, independent media producers could not benefit from a CDN at all. (Today they can subscribe to Amazon's CloudFront CDN if they can afford it.[44])

This orientation away from symmetry between users and producers later filtered down into the technologies of broadband Internet service, where it has crystallized. In 2014, wired broadband Internet across the world is provided via DSL (digital subscriber line) attached to a copper telephone network, cable modems attached to a coaxial cable television network, or a new optical-fiber network.[45] Early DSL and fiber protocols originally assumed that each user would transmit as much as he or she received. Nonetheless, by 2014 DSL protocols typically assume that the user will receive about twenty times more information than he or she transmits, cable networks assume that the user will receive three to five times more, and fiber networks assume the user will receive ten times more. Any Internet user can take an online speed test (at http://speedtest.net, for example) to reveal the decisions their Internet service provider has made about how much they may consume or produce—often labeled "downstream" versus "upstream" capacity.

When compared to the extremely constrained world of 1960s television, the Internet of 2015 must seem emancipatory: everyone has access to many more than five channels. Some forms of computer-mediated interactivity and participation are now possible, yet these are more limited than Lick had hoped. It is clear that this emerging distribution infrastructure is now strongly shaping the experience of video and the future Internet. Lick is widely acknowledged as a visionary, and it could be said that these days Lick's vision of "man-computer symbiosis" is being slowly replaced by his vision of "selective television." Yet Lick hoped that the computer-enabled television of the future would not provide everyone with access to "the same thing" and this is where Internet video departs from his aspirations for interactivity and community media production. While Lick's notion of an active audience of users producing their own media is not dead, it has been merged with the desires of traditional one-to-many broadcasters to form an interesting new technological hybrid. To produce this hybrid the Internet has often been willfully bent to train an interactive, peer-to-peer system toward the older commercial vision of "mass communication."

Certainly the Internet was originally thought to promise widespread "demassification" or "disintermediation"—anyone could be a publisher or a broadcaster with these new systems.[46] Most recent commentators on the evolution of television emphasize the significance of amateur self-publishing, noting that the Internet represents "a revolution in distribution that exponentially increases the ease of sharing video."[47] The implications of the Internet's distribution architecture are not yet clear, but they do not seem to fulfill these earlier visions and potentials. Instead, today they provide a complex, tiered system firmly biased toward large and well-capitalized media producers who have access to special networks (CDNs) and dedicated downstream bandwidth. Today it is possible to stream the Super Bowl online and post status updates to Facebook about it. We can watch *Gilligan's Island* online at a time we choose, and we can tweet about it. Nonetheless, this does not feel like the revolution Lick called for. If anything, the role of computation in today's implementation of Lick's "selective television" has been to optimize the selection of people for advertisements, not content for audiences.

The Internet Medium, Revised and Reconsidered

One important lesson from this story is that the Internet is now far from the point-to-point system of equals planned decades ago. Commentators expected that providing television via the Internet would transform television, but instead it caused the Internet's distribution architecture to become like television in significant ways. In the words of *New York Times* television critic Brian Stelter, "The Internet, which was thought to be a TV killer, is turning out to be its wingman."[48] Recent empirical studies of Internet traffic have pushed this point further, revealing that the network has reached an inflection point, where the Internet is now, for the first time, centrally organized around serving video. And this does not refer to video as a mode of communication in general, but specifically to serving a particular kind of video from a very small number of providers to large numbers of consumers.[49] The Internet is now television, or it will be soon. During peak video watching times, two providers (Netflix and YouTube) account for more than half of all Internet traffic in North America.[50] Consumer video accounted for 57 percent of all Internet data in 2013, not including peer-to-peer traffic.[51] A recent study found that at peak television viewing hours 34 percent of North American wired broadband traffic went to just one source—Netflix.[52] In another account, up to 80 percent of all network traffic during peak viewing times on one wired commercial Internet service provider went to Netflix.[53] These are not simply statistics about user preferences for video over other kinds of activity: remember that without the strenuous technological revisions to the Internet's distribution architecture described earlier, Netflix and YouTube streaming would not be possible at all.

Reflecting on the general narrative of Internet video's development, it is clear that media infrastructures do not have the essential characteristics that are often attributed to them. Just as the Internet is often thought to be "about" the long tail or user-generated content, television is often thought to be "about" liveness.[54] Jonathan Sterne countered that "the very possibilities for the experience of live television" were strongly shaped by the evolution of television's distribution infrastructure.[55] A national television distribution network was willfully called into existence in the United States before 1962 by corporate executives who were convinced that the key to profitability for the medium was advertising to a national audience. This implied that the nation must be able to watch the same television at the same time, and so AT&T was asked to construct a television rebroadcast infrastructure atop the national common carrier telephone network. Just as this chapter explained the attempts to surmount the technical challenges in distributing video over the Internet, U.S. television networks confronted the technical challenges of distributing television signals over long distances by investing in research on microwave relays and coaxial cable.

Beginning in the 1960s, engineers believed in Lick's vision, and they constructed the Internet to be the anti-television he proposed. They designed it for the people they imagined themselves to be, the reflexive users, eager to appreciate a symphony or to play in one—and not to lounge around the living room passively watching *Days of Our Lives*.[56] Even so, as the network grew and attracted the interest of commercial firms, capital eyed the Internet as a new route to profit via arbitrage. The Internet, as a new communications medium, could be a chance to displace the profitable video distribution bottlenecks of the twentieth century. Yet simply using the Internet to distribute television would not work. At first, video distribution was technologically impossible, and later the Internet's distribution infrastructure thwarted commercial attempts to develop a one-to-many video audience for almost two decades.[57] Money, resources, and ingenuity were thrown at the problem. Attempts at a solution proceeded on a dizzying number of fronts: compression, streaming, buffering, colocation, bandwidth, server farms, data centers, and others. It finally took changes to standards, protocols, and system architectures to denature the assumptions of Lick's point-to-point networking in favor of the more familiar model of mass communication as exemplified by the CDN. While the existing system is a hybrid, the direction of change has been toward a mass audience.

The key implications of this story relate to the form of content itself and the shape of our shared culture. In the United States "television" has been thought of as a container for television-specific content: a notional box that, when you look inside it, contains entertainment. The Internet is thought of as something quite distinct—a notional box that should contain something else, something different. Lick thought the box should contain symphonies and the grassroots content that users produce. Indeed, as time passed, it started to seem that his idea had prevailed. As one meme put it, the Internet is full of cats.[58] (It was a medium essentially "about" quirky, user-generated content.) To ask again a central question of media theory and a preoccupation of the Toronto School[59]— How does the container affect the form of the content it can contain?

The medium of the Internet has transformed over the last forty years from a textual system to an audiovisual one, shifting from a network of text-only emails to YouTube videos. The transformation was intentional, and not a process of maturation explained by computers and networks naturally becoming faster. Television was not just poured into the Internet box. Instead, engineers and venture capitalists worked to change the medium itself and optimize it for mass communication, providing a way to assemble large audiences for relatively few sources. These interventionists were radicals and upstarts in that they were not working for old television companies, but they were conservative in that they found that the Internet's new architecture and distribution system could not provide the older form of mass television, so they sought to revise it by looking backward for inspiration. While there was a logic at work of meeting consumer demand and satisfying customer taste, there was also a sense that the Internet user could be taught what to want, and that wanting user-produced material without commercials was not profitable. As the medium of the Internet continues to transform, it appears to be moving further from the participatory goals held by Lick and many commentators,[60] raising the question of what our normative position on access to the means of distribution should be.

Transforming the Internet medium to make television fit inside it did not simply add capabilities, making mass broadcasting easier. As the medium changed, older Internet patterns of point-to-point or peer communication were made more difficult. Today, Internet users are prohibited by their subscriber agreements from running their own servers. If a user tries anyway and becomes popular, their networking hardware no longer supports the many-to-many pattern of traffic flows that personal servers would require. Without access to a CDN, content from a mainstream, well-capitalized media company would load perceptibly faster than what the user offered, and thanks to distribution investments, traditional television content might even be seen at a higher resolution. In sum, the distribution infrastructure of the Internet has changed to make some content distribution easier and some more difficult. While user participation has not been eliminated, interactivity has been constrained to actions that surround and amplify content provided by mainstream media companies. Some of these companies, such as Netflix, are Internet upstarts, but they share strategies and technologies with mainstream media projects like Hulu (owned by NBCUniversal Television Group [Comcast], Fox Broadcasting Company, and the Disney-ABC Television Group). These video streams are not nearly the departure from *Gidget* that other writers once foresaw. The Internet is being "re-massified," but this battle is not over. Those who see a vibrant point-to-point future of videoconferencing and interactive gaming may hope to retrofit the infrastructure once again.

Distribution as Diagnostic, Distribution as Destiny

As this chapter has revealed, Internet architecture is important, but it is neither fixed nor inevitable. Internet engineers, for instance, once discussed the
trade-offs between solving the problems of video distribution via a private,
proprietary, invisible CDN (accessible to only those providers who pay for it)
and providing such facilities in public by modifying the basic protocols of the
Internet itself (making these features accessible to anyone).[61] A final assessment concerns the implication of these facts with regard to how we think about
all media systems. Investments in infrastructure make earlier decisions durable
and difficult to change, but ultimately these systems are built by people and can
be rebuilt by them. As a result, distribution architecture remains an important
site of investigation for the media scholar, as well as an avenue for intervention
by the media activist.

To the media scholar these characteristics of online video are likely amazing:
most studies of online video proceed wholly from the perspective of either the
user or the content, making the details above inaccessible. Those researchers who do consider another view often focus on industrial history or political economy, but some of these perspectives neglect either the technology or
the distribution network. Studies focusing on technology, for instance, tend to
focus on new developments in the apparatus in the home, ignoring the pipes
and wires that lead there. Much more could be learned with the distribution
infrastructure in the center of our view, echoing Sterne's calls for future analyses of the "mode of distribution" rather than production or reception.[62] In this
case, telling the story of Internet video without the above focus on distribution
could wrongly make it seem that the development of online video was purely
a matter of user preference. A future analyst might one day wrongly conclude
that the story was: "For a while, early Internet users made and watched their
own videos about cats, but then they wanted to watch mainstream media offerings like *Modern Family*." In fact, reorienting the Internet audience toward mass
offerings has been a coevolution of taste, massive infrastructural investments,
and important technological achievements.

On final reflection, such a focus on the normally invisible infrastructures
of distribution is not completely rare.[63] When the satellite emerged as a viable
technology for video distribution in the late 1960s, the transmission and distribution architecture loomed large enough to capture the attention of media
analysts of all stripes. Satellites were evocative, engendering what Lisa Parks
has called the Western fantasy of "global presence."[64] But they also seemed to
offer a reorganization of television based on transmission. Satellite signals

were naturally able to leap national borders (significant during the Cold War). Satellites also incorporated the potential for disintermediation: they could be used to establish a direct link between a source in one part of the world and a receiver in another, bypassing any local distribution networks.

Within the distribution infrastructure lies a clear picture of which speakers are valued and what content is important. The distribution infrastructure is a crucial battleground: competing visions of society are made manifest within seemingly technical struggles, yet they are also modified by the inertia of technology. Led by Herbert I. Schiller, the early critical analysis of satellites was more than an examination of the technology or political economy of a technology itself, it was a strategy for scholarly inquiry into media that focused on transmission as a critical step in the media system and the circulation of our culture.[65] To Schiller, transmission was crucially diagnostic, as it could reorganize who could speak. He emphasized over and over that communication "is dependent ultimately on some form of transmission,"[66] and that the working definition of the media and communications industries "includes data generation and transmission."[67] He wrote, "The transmission structures that are being established nationally and internationally provide . . . evidence of the character of the systems emerging in the Information Age."[68] Lick would surely have agreed.

Notes

1. For one YouTube version of this prank, see: https://www.youtube.com/watch?v=Cjbry-mObCo.

2. Susan Leigh Star, "The Ethnography of Infrastructure." *American Behavioral Scientist* 43 (1999): 377–91.

3. For an overview, see Christian Sandvig, "The Internet as Infrastructure," in *The Oxford Handbook of Internet Studies*, ed. W. Dutton (Oxford: Oxford University Press, 1999), 86–108.

4. Janet Abbate, *Inventing the Internet* (Cambridge, Mass.: MIT Press, 2000).

5. This was intended for the purpose of sharing routing and control information, not content. See J. Mogul, "Broadcasting Internet Datagrams," *IETF Network Working Group Request for Comments (RFC)*, October 1984, 919, available at http://tools.ietf.org/html/rfc919 (accessed September 20, 2014).

6. See, for example, http://www.bbc.co.uk/blogs/digitalrevolution/2009/11/rushes-sequences-vint-cerf-int.shtml (accessed September 20, 2014).

7. Laura DeNardis, *Protocol Politics: The Globalization of Internet Governance* (Cambridge, Mass.: MIT Press, 2009).

8. Allen Kent and James G. Williams, eds., *Encyclopedia of Microcomputers v. 17: Strategies in the Microprocess Industry to TCP/IP Internetworking: Concepts: Architecture: Protocols, and Tools* (New York: CRC, 1995).

9. Stepthen Segaller, *Nerds 2.0.1: A Brief History of the Internet* (New York: TV Books, 1999), 110.

10. Abbate, *Inventing the Internet*.

11. Ellen Seiter, *Television and New Media Audiences* (Oxford: Oxford University Press, 1999), 130; see also Lisa Parks, "Flexible Microcasting: Gender, Generation, and Television-Internet Convergence," in *Television after TV: Essays on a Medium in Transition*, ed. Lynn Spigel and Jan Olsson (Durham, N.C.: Duke University Press, 2004), 133–62, 138.

12. Nathan Ensmenger, *The Computer Boys Take Over: Computers, Programmers, and the Politics of Technical Expertise* (Cambridge, Mass.: MIT Press, 2012).

13. Thierry Bardini, *Bootstrapping: Douglas Engelbart, Coevolution, and the Origins of Personal Computing* (Stanford, Calif.: Stanford University Press, 2000), 108.

14. Steve G. Steinberg, "Netheads vs. Bellheads," *Wired 4.10*, October 1996.

15. Abbate, *Inventing the Internet*; Steinberg, "Netheads vs. Bellheads."

16. Bob Taylor, quoted in M. M. Waldrop, *The Dream Machine: J. C. R. Licklider and the Revolution That Made Computing Personal* (New York: Viking, 2001), 470.

17. J. C. R. Licklider, "Man-Computer Symbiosis," *IRE Transactions on Human Factors in Electronics HFE-1* (1960), 4–11.

18. It may have been ignored because he turned it in late. See the note at the bottom of the 1967 Carnegie Report, 113.

19. J. C. R. Licklider, "Televistas," in *Public Television: A Program for Action*, ed. Carnegie Commission on Educational Television (New York: Harper & Row, 1967), 201–25, 202.

20. Ibid., 209.

21. Ibid., 214–15.

22. Chris Anderson, "The Long Tail," *Wired*, October 2004, available at http://archive.wired.com/wired/archive/12.10/tail.html (accessed September 20, 2014).

23. Tim Berners-Lee, *Weaving the Web: The Original Design and Ultimate Destiny of the World Wide Web, by Its Inventor* (New York: HarperCollins, 2000).

24. Ibid., 27, 29, 42.

25. See, for example, the entry in the Jargon File: http://catb.org/jargon/html/S/slashdot-effect.html.

26. Anita Chan, "Slashdot.org" in *The Inner History of Devices*, ed. Sherry Turkle (Cambridge, Mass.: MIT Press, 2008), 125–37.

27. Daniel Dayan and Elihu Katz, *Media Events: The Live Broadcasting of History* (Cambridge, Mass.: Harvard University Press, 1994), 4.

28. For example, see Thomas Parke Hughes, *Networks of Power: Electrification in Western Society, 1880–1930* (Baltimore, Md.: Johns Hopkins University Press, 1993).

29. Licklider, "Televistas," 215.

30. DeNardis, *Protocol Politics*.

31. Stephen E. Deering, "Host Extensions for IP Multicasting," *IETF Network Working Group Request for Comments (RFC) 1112* (1989), available at http://tools.ietf.org/html/rfc1112.txt (accessed September 20, 2014).

32. Geert Lovink, *Dynamics of Critical Internet Culture: 1994–2001* (Amsterdam: Institute of Networked Cultures, 2002), 164.

33. There are many accounts of this experience online. This example is from http://community.sony.com/t5/Blu-Ray-Netflix-Online-Video/Netflix-quality-all-of-a -sudden-terrible/td-p/24363 (accessed September 20, 2014).

34. "MIT Scientists Develop New Method to Distribute Content over World Wide Web," Akamai, 1999, available at http://www.akamai.com/html/about/press/releases/1999/press_011499.html (accessed September 20, 2014).

35. J. Dilley, B. Maggs, J. Parikh, H. Prokop, R. Sitaraman, and B. Weihl, "Globally Distributed Content Delivery," IEEE Internet Computing (September/October 2002): 50–58.

36. K. Andreev, B. M. Maggs, A. Meyerson, and R. K. Sitaraman, "Designing Overlay Multicast Networks for Streaming," *Proceedings of the Fifteenth Annual ACM Symposium on Parallel Algorithms and Architectures* (SPAA), San Diego, California (2003), available at http://www.akamai.com/dl/technical_publications/DesignOverlayMulticastnewtworksforStreaming.pdf (accessed September 20, 2014).

37. Rich Miller, "Akamai Now Running 105,000 Servers," *Data Center Knowledge*, Mar. 8, 2012, available at http://www.datacenterknowledge.com/archives/2012/03/08/akamai-now-running-105000-servers (accessed September 20, 2014); Alexander Eule, "The WD-40 of the Internet," *Barrons*, February 1, 2014, available at http://online.barrons.com/article/SB50001424053111903911904579346821089194680.html?mod=googlenews_barrons (accessed September 20, 2014).

38. Eule, "WD-40 of the Internet."

39. "CDN Market Share Update December 2013" *Datanyze*, December 2013, available at http://www.datanyze.com/blog/cdn-market-update-dec-13 (accessed September 20, 2014).

40. Raymond Williams, *Television: Technology and Cultural Form* (New York: Routledge, 2003).

41. For example, https://blogs.akamai.com/2012/09/web-performance-why-one -size-doesnt-fit-all.html (accessed September 20, 2014).

42. Christian Sandvig, "The Structural Problems of the Internet for Cultural Policy" in *Critical Cyberculture Studies*, ed. D. Silver and A. Massanari (New York: New York University Press, 2006), 107–18.

43. Thomas Streeter, *Selling the Air: A Critique of the Policy of Commercial Broadcasting in the United States* (Chicago: University of Chicago Press, 1996).

44. There are also exciting new developments in CDNs, including research promising that cooperative, free CDNs might be assembled from large groups of users without a data-center infrastructure, and the advent of one CDN that offers free service to anyone (CloudFlare). It is not yet clear how these developments will unfold.

45. Or some combination of the three—for example, "FTTx."

46. Rob Kling, "Being Read in Cyberspace: Boutique and Mass Media Markets, Intermediation, and the Costs of On-line Services," *Communication Review* 1, no. 3 (1996): 297–314.

47. Amanda Lotz, *The Television Will Be Revolutionized* (New York: New York University Press, 2007), 16.

48. David Carr, "Talking Back to Your TV, Incessantly," *New York Times*, March 14, 2010, available at http://www.nytimes.com/2010/03/15/business/media/15carr.html (accessed September 20, 2014).

49. C. Labovitz, S. Lekel-Johnson, D. McPherson, J. Oberheide, and F. Jahanian, "Internet Inter-Domain Traffic" in *Proceedings of the ACM Special Interest Group on Data Communication*, SIGCOMM (2010), 79–86, 85.

50. "Global Internet Phenomena Report: 2H 2013," *Sandvine*, 2013, available at https://www.sandvine.com/downloads/general/global-internet-phenomena/2013/2h-2013-global-internet-phenomena-report.pdf (accessed September 20, 2014).

51. "Cisco Visual Networking Index: Forecast and Methodology: 2012–2017," Cisco, 2013, available at http://www.cisco.com/en/US/solutions/collateral/ns341/ns525/ns537/ns705/ns827/white_paper_c11-481360.pdf (accessed September 20, 2014).

52. Todd Spangler, "Netflix Remains King of Bandwidth Usage, While YouTube Declines," *Variety*, May 14, 2014, available at http://variety.com/2014/digital/news/netflix-youtube-bandwidth-usage-1201179643 (accessed September 20, 2014).

53. This statistic comes from a personal communication with a network operator who asked not to be named.

54. For a conceptual review of what "liveness" entails, see Paddy Scannell, *Television and the Meaning of "Live": An Enquiry into the Human Situation.* (New York: Polity, 2014).

55. Jonathan Sterne, "Television Under Construction: American Television and the Problem of Distribution 1926–1962." *Media, Culture and Society* 21, no. 3 (July 1999): 503–30.

56. The phrase "reflexive users" comes from Thierry Bardini.

57. The idea of a technology "resisting" attempts to manipulate it is taken from Bruno Latour, "When Things Strike Back: A Possible Contribution of 'Science Studies' to the Social Sciences," *British Journal of Sociology* 51, no. 1: 105–23.

58. Joel Veitch, "The Internet is Made of Cats." *Rathergood.com*, January 14, 2010, available at http://www.rathergood.com/cats (accessed September 20, 2014).

59. James W. Carey. "Harold Adams Innis and Marshall McLuhan," *Antioch Review* 27, no. 1 (1967): 5–39.

60. Henry Jenkins. *Convergence Culture: Where Old and New Media Collide* (New York: New York University Press, 2006).

61. See Sandvig, "Structural Problems."

62. Sterne, "Television Under Construction."

63. See also Nicole Starosielski, " 'Warning: Do Not Dig': Negotiating the Visibility of Critical Infrastructures," *Journal of Visual Culture* 11, no. 1 (2012): 38–57.

64. Lisa Parks, *Cultures in Orbit: Satellites and the Televisual* (Durham, N.C.: Duke University Press, 2005), 23.

65. Herbert Schiller, "Computer Systems: Power for Whom and for What?" *Journal of Communication* 28, no. 4 (1978): 184–93.

66. Ibid., 190.

67. Ibid., 185.

68. Ibid., 190.

CHAPTER 11

Consumer Electronics and the Building of an Entertainment Infrastructure

CHARLES R. ACLAND

ontrary to all the "long tail" talk of micro-audiences and narrow-band
taste formations, blockbuster entertainment remains a dominant strategy
for media industries. When *Wired* magazine editor Chris Anderson identified
the "long tail" of niche markets—a product of the distribution and consumer-
tracking precision of new computer technologies—he gave validation to entrepreneurs and social activists alike who had faith in the potential of digital
diversity.[1] The concept even provoked some prognostications about the end of
mass culture, which envisioned a digital economy driven by niche marketing.
But mass-market hits persist, they are dispersed across devices and platforms,
and they remain central to most assessments of the financial value of major
cultural enterprises. Despite Anderson's celebration of the "long tail," there
remains a powerful media strategy that concentrates on the other end of the
curve—the "fat torso."

Without question, major studios and media corporations routinely pour a
significant amount of their resources into blockbuster production, often ex-
pecting to launch or continue a franchise. Though we might be most familiar with blockbuster movies, the strategy encompasses television, publishing,
websites, apps, video games, and music. "Tentpole" films, shows, and products
are the financial and promotional centerpieces of a slate of releases, and this
"hit-driven" strategy has intensified other industrial changes. With movies, for
instance, the contemporary blockbuster era is characterized by the expansion

of what Thomas Schatz has called "conglomerate Hollywood," in which films are produced and distributed by global multi-industrial concerns, and figure prominently in the promotion of cross-media products.[2] For example, a movie can feature a recording artist and a soundtrack, reference a current television show, and include product placement for a clothing line and a soft drink, each association derived from ownership advantages or contractual arrangements. Coincidentally, lucrative ancillary markets and windows of exhibition prompt the coordination of works across television, home video, computers, and electronic devices. This coordination extends to cross-promotional efforts between content and hardware. So just as movies signal their own mutating commodity form—for instance, as ring tone, as DVD, as TV, and as homepage—they can also promote the conditions and materials necessary for their circulation, including the projector, the smart phone, the tablet, the DVD player, and the monitor.

In light of this cross-media industrial circumstance, the highly visible, international, big-budget blockbuster production makes manifest the developing relationships among media forms. The blockbuster, in a time of expanding talk and exploitation of "long tail" microcultural economies, advances multiple products and devices at once, and it does so through the formal mechanisms of cross-media promotional deals as well as through indirect support by being the most highly valued content for various platforms. The "technological tentpole"—namely, that feature property or franchise entry that strategically promotes cross-media commodities along with new generations of devices, platforms, and hardware—has become notably prominent in the American media business. In other words, blockbusters are not just films, television shows, and music; they are vehicles for the advancement of the broader technological, cultural, and economic system.[3] This system, today, consists of and depends upon an everyday landscape of electronic devices, a landscape that is so well established that it serves as an infrastructure for the distribution and consumption of media materials. Though unevenly accessible, and as provisional as its current manifestation surely is, this dispersed network of devices forms an entertainment and informational infrastructure upon which dominant cultural and economic practices transpire.

This chapter explores the intersection between entertainment and digital technology, between audiovisual content and the world of consumer electronics that has built an infrastructure of devices. Moreover, this chapter takes up the entertainment technology industry event as a featured location at which ideas about the relations among content and electronics are being explored and which organizes a consumer electronics infrastructure for entertainment and information. Take, for example, one venue that addressed the entwinement of entertainment and technology, the 2010 3D Entertainment Summit in Los Angeles. This two-day conference included keynotes from DreamWorks Animation SKG's Jeffrey Katzenberg, blockbuster auteur M. Night Shyamalan, Sony Pictures Technologies president Chris Cookson, and AEG Network LIVE president John Rubey. The event provided a networking opportunity for producers, studio executives, theatrical representatives, consumer electronics manufacturers, financial analysts, and legal practitioners. The event sponsors mirrored the same cross-section of technology and media concerns: Sony, Mitsubishi, Technicolor, JVC, IMAX, Dolby, Panavision, DreamWorks, DPL Cinema Texas Instruments, and the Consumer Electronics Association, to name about half of them. Panels and talks addressed 3D's financial and audience prospects for movies, television, mobile devices, games, and advertising, and a prominent theme across all was convergence and synergy, whether in transmedia storytelling, promotion, or business models: "3D Games Capitalize on Hollywood Success," "Almost Nothing Will Have the Impact on 3D Sales as Sports," and "How 3D Games Will Drive Consumers to Upgrade to 3D TVs and Gadgets."[4] The ambitions for technological tentpoles are not always so explicitly stated— here, how select blockbuster games drive hardware upgrades—but certainly, at an event such as this one, the joint development of the device and content industries, along with presumptions about the future of audience activity, was evident.

Transmedia, spreadable media, social media, second screen, story worlds— these are some of the current go-to terms designating the media environment of popular entertainment and informational services. These monikers encompass what used to be called simply "television" or "the movies."[5] Together they attempt to designate the high degree of mobility and mutability of cultural materials, seeing works as being stretched, squeezed, and supplemented, and proposing that an unstoppable textual surplus spills beyond the boundaries of any single medium or any single authorial design. Moreover, "transmedia" implies that this textual surplus should be attended to and exploited by producers of media content, electronics designers, and software developers. The speed of acceptance of variously named ideas about contemporary cross-media practices is astonishing. "Second screen," for instance, denoting the smart phones and tablets that audiences consult while watching a primary screen, typically the television set, sits solidly as part of industry lingo. Though it first appeared in 2010, the trade press represents the concept as so timelessly obvious that it now treats the idea as a little behind the times, perhaps even old hat.

The appearance of an identifiable lingo points to a tacit agreement about the analytical and explanatory advantage terms offer. The impression of a rapidly settled consensus about the necessary direction of industrial and technological development is a sign of cultural shifts and emergent practices in formation. Bear in mind, the relationship to actual cultural practice is tenuous at best. The relative availability of "transmedia," "second screen," and other such terms tells us that an emerging common sense about cross-media relations is in process. These terms, and their associated rationales and illustrations, provide ways to build and imagine practice, community, entertainment, and news. Their availability as explanatory tools may have an impact upon how people engage with media, how people understand their engagement, and what expectations are held for new content and devices, even though "people" only designates a select cohort of electronic opinion leaders we might call *the wired class*. And as concepts are strategically taken up and operationalized by media entrepreneurs, they serve up ways to envisage and develop freshly exploitable revenue streams.

The larger development is that an emergent common sense provisionally settles social and cultural protocols for intermedia relations. The moments of textual transfer and flow, the conditions of technological assembly, the synchronization of devices, lateral transport of content between devices, and the updating and backward compatibility of software are just some of the material practices that connect the things we call readers, viewers, players, and browsers. One result has been that media platforms and devices have an accented role in industrial strategy and a higher visibility for audiences. Our cross-media era has produced platform plenitude. Navigating this plenitude has become a necessary condition of daily life for that wired class of people living in contemporary advanced technological societies, and it has made technological reflexivity a prominent feature of entertainment and information media. Spectators regularly engage in *branded viewing experiences*. We attend an IMAX movie, a RealD 3D presentation, and a Dolby Atmos sound environment.[6] One does not watch a movie on an iPad. One watches an iPad showing a movie.

The branded sphere of abundant "tech specs" puts popular entertainment in conversation with the industry that promulgates and builds the world of evolving platforms and devices, namely consumer electronics. I do not want to neglect all of the communities and sensations that can be activated by any media form. And I am not returning to some precultural economic bottom line that answers all. But let's also not pretend to be talking about something more ethereal and wondrous than it is. Forget the fetishism of the smooth in Apple's design template or the implied mystery of Xbox. It is consumer electronics. It is appliances—appliances with aspirations, sure, but appliances all the same: think of them as fridges whose crisper contents can be shared with other fridges, toasters that cook prepackaged toast, and vacuum cleaners that archive precious family dust. Curiously, celebrants represent "the Internet of things" as a literal manifestation of these fanciful exchanges, when in fact it is largely a developing network for controlling and gathering data from electronic appliances. While there are varying definitions and boundaries to what exactly this hybrid entertainment-information-technology industry is and will be, I strategically emphasize the consumer electronics business in part because even the name has a clunkiness to it. It is an inelegant term that conjures images of cardboard boxes, packing material, multilingual instruction booklets, batteries, and AC plugs. It deflates the promises of smooth convergences and imagined immateriality, of swarms and hives and clouds.

Sometimes deflation is what we need. In the boom and bust cycle of scholarly attention, an inflationary period has produced high-minded proposals about the digitally networked universe. But the lived world of fun, work, and things deserves a better account of the conditions that occupy our thoughts and homes, one that includes structuring, facilitating, and inhibiting features. The best of the production-culture literature is doing this deflationary work, revealing the unglamorous endeavors that are the actuality of media-making—for example, Vicki Mayer's first chapter of *Below the Line*, about electronics assembly in Brazil.[7] Accounts of the environmental catastrophe of our screen culture also return us to the limits of industrial processes and consumer appetites, as represented by Richard Maxwell and Toby Miller's work.[8] Lisa Parks describes turns such as these as "thinking elementally" about media, keeping resources and materials front and center in our analyses.[9] Putting consumer electronics into the scholarly purview of media scholarship helps challenge the naturalism that creeps into accounts of new media flows. Doing so illuminates the concerted and elaborate efforts to build our infrastructural environment of audiovisual components and to establish a common sense about the interrelationship between audiovisual content producers, electronic devices, and platforms that litter our lives.[10]

"Platform," here, describes the contours, protocols, and generations of devices through which we engage with cultural materials. Electronic devices may vary in size, color, design, price, and so on, but each represents a specialized association with particular platforms, which systematizes the relation between hardware and software, including the selective flow of content between devices. In film parlance, "platform" refers to a staggered and gradual releasing pattern for a film, as opposed to wide or saturation releasing, so the term is partially tied to a particular exhibition strategy. Gaming, though, advances a connection to hardware and compatibility with games and editions. Platform, in this way, presents as a location at which something can appear and as a set of limitations upon that appearance; it is a staging ground, a host environment, and a setup for something else to transpire. Tarleton Gillespie outlines this understanding of platform as consisting of an ideology of "open, neutral, egalitarian and progressive support for activity."[11] He explains that prominent contemporary usage of the term highlights the intermediate elements of media systems. Platforms are not neutral stages upon which interactions transpire; they are deeply embedded technological structures that both facilitate and inhibit communication. He emphasizes the discursive function: "A term like 'platform' does not drop from the sky, or emerge in some organic, unfettered way from the public discussion. It is drawn from the available cultural vocabulary by stakeholders with specific aims, and carefully massaged so as to have particular resonance for particular audiences inside particular discourses." Further, he says, "[T]hey represent an attempt to establish the very criteria by which these technologies will be judged, built directly into the terms by which we know them."[12] In our context, these criteria inflect platforming with *interchangeability and a sense of relation among media*. These connotations highlight the way content is variously shaped and shared by platforms that deliver, present, archive, and circulate films, shows, music, and games. Thus, "platform" resides alongside other terms capturing the inherent instability of what we think of as a medium, like "convergence," "intermedial," "transmedia," and "format." These terms are not equivalent, and they emerge out of different debates, but they do share recognition of the conceptual limits of medium specificity.

As lives are ever more cluttered with various generations of gear and gadgets, we also confront more insistent images of frictionless movement among platforms and textual variations. The decisions to advance transmedia productions, as we see with American blockbuster franchises, take for granted the evolving adoption of multiple devices and the intensification of itinerant usage. The transmedia blockbuster rests upon a landscape of dispersed screens, players, smart phones, tablets, and monitors, and in this respect consumer electronics functions as an infrastructural element for the digital media industries. José van Dijck observes, "As a result of the interconnection of platforms, a new infrastructure emerged: an ecosystem of connective media with a few large and many small players."[13] Her work, in which she documents the movement from participatory online networks to a corporatized culture of connectivity, links the microsystems of platforms to larger "*techno-cultural constructs* and *socioeconomic structures*."[14] In doing so, van Dijck understands the dominance of Web 2.0 and its various platformed iterations as having matured into a "functional infrastructure."[15]

Van Dijck's platforms refer to Facebook, YouTube, Twitter, and the like, where my use here encompasses different generations of hardware designed to accommodate particular forms of media content. But consistent across these definitions is the focus on their infrastructural presence and operation. As elaborated in infrastructural analysis, a foundational feature of media industries is the sculpting of the media environment, a function that has as much to do with technological firewalls between media as it does associations permitted by technological frames and sanctioned in practice.[16] Geoffrey Bowker and his co-authors have studied "information infrastructures," understanding "infrastructure" as *"pervasive enabling resources in network form."*[17] For "cyberinfrastructures," they take this to include not only the technologies themselves but also the conceptualizations of problems and research agendas, as well as the circulation of results. This resonates with Susan Leigh Star's elaboration of infrastructural features: it is embedded in structures and social arrangements; it appears transparent but becomes visible when service ruptures; it is installed, with a degree of permanence, highly complex, and cannot be overhauled at once; and it is taken for granted by users, involves standards and conventions of practice.[18] Living as many do among a harvest of screens, keyboards, remotes, and consoles, these features describe a consumer electronics infrastructure, which is today a "pervasive enabling resource in network form."

Webs of connectivity among devices and their related protocols are simultaneously fences between media. Prior to some ultimate, and ever-receding, horizon of fully integrated media, as Lisa Gitelman has smartly alerted us, we exist in a context of technologically, administratively, economically, and culturally sanctioned protocols for practice and intermedia relations.[19] Or, as Gillespie puts it, contra the transmedia celebrants, "'Platforms' are more like traditional media than they care to admit."[20] Accordingly, setting conditions for "transmedia" platforms—as a goal, a possibility, an inevitability, and a future—has not been the result of fresh technological capabilities alone but of how we understand, explain, and take up selectively featured capabilities. So how, indeed, was this entertainment infrastructure built? How did the consumer electronics business become more tightly integrated with the entertainment industry?

Consumer electronics is a giant of an industry. It was a roughly one-trillion-dollar business in 2011, with U.S. spending making up $190 billion of that amount.[21] For the United States, this was only $60 billion a decade earlier in 2000.[22] "Hollywood versus Silicon Valley" has become a clichéd way to understand the competing stakes of each business domain. This narrative serves as the primary focus for J. D. Lasica's *Darknet*, which chronicles the ways Hollywood has resisted the technological innovations of the high-tech business in favor of more control over the circulation and use of their properties.[23] But this story, with its impression that the realm of entertainment is naturally at odds with IT and electronics, obscures the substantial work done to align each sector to their mutual benefit.

One might rightly point out that consumer electronics in one way or another has always been part of the U.S. entertainment industry. The long view certainly stretches back to the early Hollywood investments of RCA and Westinghouse. But today electronics is, in many ways, *the* story, that is, the development of new technology—for production, distribution, consumption, narrative expansion, character monetization, and so on—is a dominant facet of financial investment, talent exploitation, and audience navigation. Reflecting this turn, in 2008 *Variety* began to honor influential technology developers and promoters with its annual "10 Technology Innovators to Watch" list. Recent reconfigurations of the websites for *Variety* and the *Hollywood Reporter* have given more prominent attention to consumer electronics devices and IT gear, in addition to expanded coverage of the digital effects business. In fact, the revamping of *Variety* in 2013 following a change in ownership made the publication far more attentive to digital media than it had been in the past, with the weekly edition adding a regular page by Marc Graser, called "Executech," profiling the newest electronics devices, accessories, and apps. And in the same way that ComicCon, the annual comic convention held in San Diego, has become a major launching pad for Hollywood productions over the last decade, so too are various gaming and electronics gatherings now taken as suitable venues for promotional pitches for new theatrical releases and other entertainment products.

Direct turning points toward an articulation of entertainment with consumer electronics were Sony's purchase of Columbia in 1989 and Matsushita's acquisition of MCA/Universal in 1991. Both Sony and Matsushita (better known to American consumers by its brand Panasonic), as large international Japanese technology corporations, were leading manufacturers of VCRs. Business historian Sea-Jin Chang observed that Sony's strengthening interest in securing production lines and content providers emerged from their understanding that their Betamax would not have lost the VCR format battle to VHS if they had had a library of filmed material to back their format.[24] Whatever the rationale, many industry analysts and prognosticators interpreted these acquisitions as a logical fit between hardware and software. Sony's and Matsushita's high-profile, headline-generating purchases helped provide evidence for claims of "synergy" between electronics manufacturers and entertainment producers. Despite the fact that Matsushita's stake was short lived and that they divested themselves of MCA/Universal in 1995, the language and logic of synergy continued.[25]

This logic expounded a claim that media were evolving and that their DNA was being gradually modified to facilitate the convergence of all media. A Darwinian natural-selection motif intensified as digital data became an accepted "degree zero" for all content. Obfuscated in this discourse of the natural life of technological progress were the concrete decisions made to permit such media integration. Prior to 1996 it was not a given that DVDs would be able to play on computers or that the eventual adoption of restrictive coding for different geographical regions for DVDs would not also apply to other devices.[26] But the ascendant logic of convergence helped open the closer movement between content providers, high-tech innovators, and electronics manufacturers with the equal compatibility of DVDs with DVD players and personal computers, and the later expansion to streaming, digital downloads, and virtual lockers.

Ownership structure is only one way to chart the developing relations among entertainment industries and consumer electronics. The very notion that connection should be made—that relations among media content and devices should be more elaborately entwined—is the product of concerted intellectual efforts by trade organizations and business leaders. There are many venues where this force is exerted, including universities and financial institutions. But one especially distinctive stage has been industry summits and conferences at which representatives from seemingly disparate industrial concerns are brought together to exchange information, to present new developments, to forge business relationships, and to debate future market conditions. In recent years, even a casual reader of *Variety* and other entertainment trade publications would have noticed the explosion of events that focus on technological alignments between content producers and devices manufacturers. An effort to construct an inventory of these joint entertainment and technology industry events, using notices in trade publications and corporate press releases, shows one taking place in 2000, two in 2006, and sixteen in 2012.

The industrial summit is an important "contact zone" between actors in relatively powerful corporate positions and a trade public, as John Caldwell has elaborated.[27] His research has dealt with production-oriented trade shows, concluding they involve attempts "to . . . 'fix' . . . common identities and amorphous institutional borders through symbolic means."[28] Caldwell writes that the gatherings function "as industrial consensus-forming gatherings; as group self-reflection activities; as cooperative negotiations responding to new technology threats or economic changes; and as socio-professional networking rituals."[29] All these features, and especially the latter, are echoed in the confabs sampled here, but I want to highlight the notable cross-sector programming involved, one that combines studio executives, producers who are seen as technological auteurs, and spokespeople from electronics and IT companies large and small. The result is that an informal umbrella category of "entertainment technology" has emerged to accommodate a range of enterprises, providing the conceptual armature necessary for an entertainment infrastructure of consumer electronics.

The entertainment technology event speaks simultaneously to film, television, gaming, and online producers, as well as device, hardware, and app creators. Examples are DICE Summit—Design, Innovate, Communicate, Entertain—begun in 2001, part of the Academy of Interactive Arts and Sciences, founded in 1996; Electronic Entertainment Expo (E3), founded in 1995 by the video-game trade organization Interactive Digital Software Association, which was renamed Entertainment Software Association in 2003; and the 2010 launching of Future of Film Summit, Hollywood IT Summit, and the Entertainment Apps Summit. The affection for the moniker "summit" lends a diplomatic or détente aura to the proceedings, as though these are exercises in peacemaking among traditional combatants. And surveying the abundant paratextual documentation each of these events produces—including schedules of talks, promotional material, keynote transcripts, press releases, online video postings of presentations, and trade reporting—the term "lovefest" is not out of place. The entertainment technology event is not really a place to debate; it is a place to affirm and organize ideas and strategies appropriate to developing markets, to express participation in that development, and to imagine cross-media industrial stability, even as "flux," "innovation," and "game-changing" appear as dominant topics and expectations.

The people delivering keynote presentations are business leaders who headline a schedule of panels addressing technologically driven changes to the media entertainment business. For example, the 2009 3D Entertainment Summit's keynotes were Jeffrey Katzenberg (DreamWorks), Michael Lewis (RealD), Sandy Climan (3Ality Digital), and Peter Bart (*Variety*). Katzenberg is frequently invited to such events and is a sort of idol of entertainment technology development. In effect, these meetings circulate a hierarchy of authority in which business success is seen as evidence of innovation and can result in prominent speaking placement. Attendance varies from several hundred to several thousand, meaning there is a good deal of activity beyond the star speakers. Summits are quasi-public venues for exchanges and developing relations, and they are designed to prompt further interactions and collaborations by bringing people together who previously would have been in distinct orbits. They involve trade and, on occasion, general press coverage, but they also build momentum and set agendas. Each iteration assembles what industry agents are talking about *and* what industry agents *should* be talking about. A prevailing sense of privileged access to "what's next" pervades the panels and presentations. The meetings are not the determining forces for convergence, but they are one facet in the regularization of who speaks to whom, when that speaking transpires, what the primary agenda points are, and, primarily important, how that intersection is propelled into a seasonal and predictable trade news story, one that can jump to general audience periodicals. In addition to the business deals and arrangements that emerge from meetings, the story of entertainment and technology summits is that *there are these summits*, which makes manifest this newly settled genre of cross-media networking, prognosticating, and promoting. Their relative regularity produces the discursive conditions for building and fortifying an audiovisual infrastructure.

For instance, *Variety* and Digital Hollywood Inc. launched the Entertainment and Technology Summit in 2010, which has been held regularly since, describing it as an "intersection of content, technology, and entertainment."[30] Their press material indicated that the event was for "social media campaigns for advertisers and marketers; successful programming across apps, mobile, online and other platforms; consumer electronics' eager hunt for film/TV content; and technology trends in blockbuster film production."[31] The 2011 edition of the summit included a slate of featured speakers that displayed a decidedly cross-media investment: Steve Mosko (president, Sony Pictures Television), Thomas Gewecke (president, Warner Bros. Digital Distribution), Robert Kyncl in charge of Google's film and TV division, and producer-director Eli Roth.[32]

Variety's involvement in effectively producing events they report on can be expected to intensify as a result of its acquisition in 2012 by Penske Media Group, the online news organization that runs Deadline.com and Movieline.com. But Digital Hollywood Inc. has a special place in the regularization of such trade events. It is an enterprise whose sole purpose is to organize summits and conferences that bring new media techno-tycoons and tycoon aspirants into contact with entertainment producers. Digital Hollywood is an outfit that coordinates dialogue between, and produces networks for, the entertainment business and the electronics/computing business. Founded by Victor Harwood in 1990, the company grew out of his earlier Multimedia Expo, from 1987. According to their promotional material, their regular events—the Media Summit New York, Building Blocks, Advertising 2.0, and Digital Hollywood Europe—offer insight into new directions in technology development, into consumer engagement with technology, and into trends worthy of financial investment.[33]

Digital Hollywood is not the only business operating in the entertainment technology summit-planning field. Unicomm, with the Bob Dowling Group, organizes trade shows, and has worked jointly with *Variety* on the 3D Entertainment Summit, the 3D Gaming Summit, and the MultiScreen Summit.[34] Other industries have similar initiatives. For example, in the advertising industry, BBDO runs Digital Lab "to raise and maintain the 'Digital IQ'" of its employees and clients, and where they typically provide extensive coverage of the International Consumer Electronics Show. Digital Hollywood and the other similar concerns likewise understand their operations as educational, producing venues for the relay of fresh information and ideas about developing technology and shaking up previously stable modes of business decision making. To this end, they create networking occasions as well as circuits upon which credible voices about pressing tech topics are identified. Some of the people Digital Hollywood Inc. has placed in these events are Intel's Donald Whiteside, Electronic Arts' Bing Gordon, producer Dick Wolf, HDNet's Mark Cuban, as well as Sean Combs, Courtney Love, and Carl Bernstein.[35] Celebrity status and the position held in an entertainment/technology company secure a place on the schedule, and a place on the schedule builds capital as an interpreter of new tech trends. As Pierre Bourdieu's work has proposed, critical attention to position-taking among cultural agents moves us away from concepts of the charismatic creator or critic toward a field of forces producing winners and losers in a game of reputational and economic value.[36] Such is the case with the featured participants in the circuit of entertainment technology summits. Making a bid for his own status as a visionary business analyst, Harwood claims to have coined the phrase "Digital Revolution," an assertion that is somewhat difficult to verify or disprove. Of the firm's many sponsors, the major "associate" sponsor for Digital Hollywood is the Consumer Electronics Association (CEA), the primary trade lobby for the electronics industry.

Table 11.1 in the appendix presents a selected chronology of entertainment and technology industry events run by Digital Hollywood and others. While the events listed concentrate on those taking place in California and Nevada, comparable summits and symposiums have transpired worldwide. Though incomplete, it provides a reference guide for this developing discursive structure. The atmosphere of information exchange focuses on entrepreneurs and producers who relate experiences and impressions of how they have exploited, or plan to exploit, new devices, delivery systems, and technological processes. Topics veer between "how to" to "what is." They involve 3D, mobile devices, games, Web content, social media, and smart TVs, among others, providing a portrait of the expanded concerns of Hollywood as that of a reliably evident electronic infrastructure for delivery paths, convenience-seeking audiences, mobile niches, and multiple-device owners and users. Sponsors are many and varied, but one does see several particularly active organizations represented across these events, including *Variety*, Sony, Pricewaterhouse Coopers, RealD, thismoment, and Cinram. Few, though, are as consistently prominent as the CEA. Among their event investments are several iterations of the Entertainment and Technology Summit, the Future of Film Summit, the Film Marketing

Summit, the Entertainment Apps Conference, and the 2nd Screen Summit. And their premiere trade convention, the annual International Consumer Electronics Show, has sponsored the Entertainment Protection Summit, August 2010, and the Producers Guild of America's Produced By Conference, which was held in June 2012 at the Sony Pictures Studio in Culver City, as well as hosted Entertainment Matters.[37]

The CEA has been operating since 1924, but it has grown rapidly over the last decade; in 2000 it represented six hundred companies, but by 2012 that number was two thousand.[38] The CEA runs its enormous trade show, the International Consumer Electronics Show (CES), annually in January. The International CES began in 1967 in New York City and settled in Las Vegas in 1996, strategically situated closer to the heart of the American media entertainment business in Los Angeles. This event has been a favorite venue for the launch of new electronic products, notably the VCR (1970), laserdisc player (1974), CD and camcorder (1981), DVD (1996), PVR (1999), 3D HDTV (2009), and electronic tablets (2010). Its popularity is such that every year, a wave of trade and general periodical writers scour the convention floor and corporate booths for the next essential consumer electronic commodity and to take stock of this industry that has so thoroughly become part of the texture of daily existence.

Figure 11.1. Convention floor of the 2011 International Consumer Electronics Show, Las Vegas. Photo by Justin Sullivan/Getty Images News, Getty Images.

The International CES had not been a prominent stop for American entertainment companies. This began to change in 2001, when the CES offered a three-day track of panels and presentations called "Digital Hollywood," not to be confused with the aforementioned company of the same name.[39] Described as a response to the pressures of digital convergence, the "Digital Hollywood" sessions addressed new multimedia markets and delivery capabilities pertinent to TV, film, publishing, cable, and Internet firms. While structured to draw companies in distinct sectors together, it was also billed as an "educational" event, meaning an objective was to inform entrepreneurs of developing trends in consumer electronics.[40]

Prominent CES participant Sony has been key in forging relations between consumer electronics and the entertainment business. Since their joint venture in CBS Records in 1968, which they eventually acquired outright in 1988, Sony has experimented with expanding its operations from electronics to embrace entertainment content. Introduced in 1996, Sony's VAIO (video audio integrated operation) was a multimedia home computer designed to advance the integration of AV and IT, further committing their growth to blurred boundaries among consumer media devices. Sony continued this when they introduced game platform PlayStation 2 with DVD capabilities in 2000, and then again in 2006 when they launched the Blu-ray compatible PlayStation 3.[41] That same year, CEO Howard Stringer, freshly appointed as the first non-Japanese head of Sony, delivered his "visionary" keynote at CES, describing the four pillars of his company's strategy: e-entertainment, digital cinema, higher definition, and PlayStation. As Stringer put it, "Content and technology are strange bedfellows, but we are joined together. Sometimes we misunderstand each other, but isn't that, after all, the very definition of marriage?"[42] The obvious hawking of Sony products in this ostensibly learned assessment of the future of consumer economics was perfectly in step with the venue and the objectives of the CES. Of broad importance to the entire sector, Stringer reiterated the claim that convergence of media and content was a necessary feature of the development of these industries. As part of his presentation, and to help introduce the Sony Reader, Stringer brought out Dan Brown, author of *The Da Vinci Code*, who praised the development of e-books. Following Brown were Ron Howard and Brian Grazer, director and producer (respectively) of Sony's film version of *The Da Vinci Code*, and the film's star Tom Hanks, who talked about digital cinema and Sony's digital cameras. CBS sports announcer Greg Gumbel then came out to talk to Stringer about Sony's HD televisions, after which Michael Dell, of Dell computers, hit the stage to confirm the importance of Sony's Blu-ray to the information-technology sector.[43] Supplemented with film clips and product demonstrations, the performance blended the promotion of content, devices, and IT.

The integration intensified in 2007 when Sony decided to forgo attending NATPE—National Association of Television Programming Executives—where it was known for its particularly elaborate and glitzy presence, instead setting up a comparable display at CES. In doing so, they featured star appearances, in this instance bringing Jerry Seinfeld and Tony Bennett along to perform.[44] This was the first time a major had done this, and it has been followed by copycat bookings in the subsequent years by other corporate participants. Incidentally, seeing the growing interest in the hardware side of the business, NAPTE responded with a counterposition for its conferences, adopting the slogan "Content First" in 2010.

Variations on "Hollywood Wired for Business" are now familiar banners for articles during CES that espouse the gold-rush fervor of new media industrial economics.[45] And yet for all the sophisticated ways entertainment-technology champions declare the efforts to cling onto old analogue models of business as ruinous, and as much as they see categories like distribution and franchising as being supplanted definitively by hive sourcing and story worlds, there remains a distinctly conventional tone to the way they still invoke a division between consumer electronics providing hardware and entertainment producers providing software. For instance, it has been noted that "every January, the major electronics manufacturers trek to Las Vegas to unveil their latest consumer gadgets, and for Hollywood that's meant new platforms on which to distribute its films, TV shows, music and games, along with other content. The major entertainment players have caught onto that, with the last two years seeing a surge in registered attendees from the biz."[46] In registering the rising participation of people from "the biz," recognizable media categories appear, not novel hybrids. Regardless, these events crucially take for granted the existence and plenitude of disparate electronic devices loaded with software applications that receive, display, and make available the designated content materials. The pressure to "innovate," to build a "game-changing" transmedia work, and to launch a "killer app" all rely on a foundational new audiovisual infrastructure, that is, a network of powered gadgets and platforms.

Capitalizing on Hollywood's rising participation in consumer electronics, *Variety* and Cricket Wireless launched Entertainment Matters in 2011 at CES. As longstanding CES's senior vice president of Events and Conferences Karen Chupka said, this stream of programming was "to customize the CES experience for the entertainment community so they can find the right exhibitors, attend targeted conference sessions and network with the crucial business partners."[47]

That year, it was coproduced by United Talent Agency and reportedly had more than nine thousand participants.[48] The following year, Entertainment Matters 2012 had actor Eliza Dushku as its "ambassador,"[49] when attendance reportedly soared to thirty-two thousand.[50] With Entertainment Matters, as the trade reporting emphasized, CES was "not just a hardware show."[51] In fact, so eager was the CES to foreground this, their press releases stated that the name of the event was *not* "Consumer Electronics Show" but "International CES," though subsequent reference could be to "CES."[52]

Variety continued its sponsorship of Entertainment Matters for the January 2013 gathering, but this time with an expanded presence. In addition to the Entertainment Matters track, they hosted two one-day summits. One day's theme was "Film and Technology" and the next was "Television Programming and Technology," again displaying a predictable split that belied all the surrounding talk about the disappearance of the industrial divisions. Additionally, they devoted special editions of the magazine to each day of these events, called *Entertainment Matters Daily*.[53] After years of smoothing the intersection between devices, platforms, and content, the idea of convergence evidently remains novel. Or, alternately, the conventions of trade reporting and promotion require painting habitually covered issues as though they are fresh and new. We might call this tendency *innovation inflation*. In a fine example of this, a press release for Entertainment Matters 2013 summoned an impression of a courtship dance between entertainment, consumer electronics, and IT: "Getting to know you."[54] Note, this release appeared not in 1993 nor even in 2003, but in 2013. Similarly, the release's description of convergence might have been written at any point during the last two decades: "TV set makers want to reach the same audiences as TV show producers. Same goes for the people who create music and the companies that make devices to store and play back audio recordings. In the newer videogame industry, the connection between hardware and content is often closer."[55] Further testifying to the essential blending of sectors, Scott L. Brown, senior vice president of Technology and Strategic Relations at Nielsen, stated that CES is "the directional compass for the confluence of technology, innovation and the media experience." He continued: "No other show provides such a rich networking opportunity, maximum exposure to breathtaking technology advances and exceptional value afforded to media entertainment attendees."[56] At the 2013 edition of Entertainment Matters, Tim Kring, executive producer of *Touch* (FOX) and *Heroes* (NBC) and a popular presence at entertainment technology summits, presented a keynote on cross-platform storytelling.[57] Providing additional evidence of new categories of converged media, the 2013 event also featured a track of panels on Content and Disruptive Technologies, which scheduled sessions on cloud privacy and security, ubiquitous content accessibility, TV streaming delivery, cloud-friendly hardware, and second-screen usage by "media stackers."[58] Attendees repeatedly praise CES as an influential venue for linking content and technology, and for providing valuable networking opportunities for industry sectors that are not habitually in the same orbit. CES involves showcasing new products and prototypes of products that might be viable in the near future. But its function as an opportunity to strike alignments between technology and content is paramount.

The constructed proximate relationship between consumer electronics and the entertainment business is additionally manifest in the Home Entertainment Hall of Fame. *Video Business*, the leading weekly home entertainment trade publication, ran from 1981 to 2010. It founded and hosted the Home Entertainment Hall of Fame from the start of its operations to recognize business luminaries who had a vanguard role in the expansion of the home market for video. Reed Business Information owned this publication. Reed was also the parent company of *Variety* from 1987 until 2012, at which point it sold the famed entertainment trade to Penske Media Group for $25 million. Just prior to this latest sale, when *Video Business* folded, *Variety* took the helm of the Home Entertainment

Figure 11.2. Director Barry Sonnenfeld showcasing *Men in Black 3* with *Variety* writer Marc Graser at the 2012 International CES. Photo by Bryan Steffy/WireImage, Getty Images.

Hall of Fame ceremony in 2010, at that time honoring Netflix executive Ted Sa-
randos, Anchor Bay Entertainment president Bill Clark, and the Blu-ray Disc
Association, considered responsible for orchestrating the acceptance of Blu-ray
as the new home-format standard.[59] The 2011 edition of the newly titled *Variety*
Home Entertainment Hall of Fame honorees again displayed the particular
formation of entertainment technology advanced by the summits: the Harry
Potter franchise, Nintendo, John Marmaduke of media retail chain Hastings
Entertainment, CEO of Best Buy Brian Dunn, and, completing the circularity
of accolades, Gary Shapiro, president and CEO of the Consumer Electronics
Association.

The entertainment technology events described in this chapter are not the
high-pressure video "bake-offs" Caldwell describes so well in his research,
and I doubt they are as directly exploitive in the same way (though the promi-
nence of "booth babes" at CES has become a point of protest, especially criticized by some of the younger video game companies who have become attuned
to sexist representational practices). But they are similar as spaces for public
interaction and disclosure, which can be informational, promotional, and
reputational. The talks and topics affirm the certainty of new relations among
media and platforms, and a teleological trap is performed year after year, one
that builds an infrastructure of consumer electronics even as it treats it as an
existing foundational hardware structure. Take, for example, an advertisement for the first CES Entertainment Matters in 2011.[60] One featured panel
presents "Hollywood Creative Masters" Conrad Green, Gale Anne Hurd, Tim
Kring, Jeff Ross, and Tom McGrath, all of whom are "top content creators . . .
offering insight into how to take advantage of the new opportunities open
to them through social networking, transmedia, mobile platforms and other
technologies." The other featured panel is "Tech Tussles," where corporate
executives address "technologies that can excite and frighten the entertainment industry." These sessions presume a state of the industry that is both
reliably extant and has yet to unfold fully. As evident in other entertainment
technology events, these sessions maintain a sense of impending industrial
and market upheaval while simultaneously reassuring that there are ways to
exploit that upheaval. The future of entertainment technology excites and
frightens; it induces expectation and anxiety.

The details provided in this chapter are a sketch of an intricate arena of networking, knowledge sharing, position taking, and category constructing about
an entertainment infrastructure upon which new industrial practices rely. As
treated here, the entertainment technology event gives us a window into the intense conceptual and financial investment in platforms, devices, and intermedia

initiatives that exists among electronics, IT, and media corporations alike. The summits are part of the backstory to the rise of the "technological tentpole" and of an economic model based on branded viewing experiences. They do so, in part, by recognizing producers of "technological tentpoles," including James Cameron, Gale Anne Hurd, and Tom Kring, alongside equipment manufacturers and software developers. The entertainment technology events are venues at which an industrial common sense about the business of audiovisual materials is eked out, extended, and fortified. In the process, the culturally and industrially dominant position of gear and gadgets is further normalized and settled. Insomuch as consumer electronics and information technology form a basic web of media circulation and connection, the field of the entertainment technology event is a regularized discursive engine that builds and advocates for an audiovisual infrastructure.

In the dense fog of myths and presumptions about technological innovation, and in light of the standard corporate drive to identify and exploit lucrative veins of consumer desire, the "educational" trade venue serves an evermore crucial function. It helps orient participants and build agreements about what developments are pending and who is most apt to understand them. Trade conventions and summits are a trail of efforts to stabilize a space for industry segments to inhabit, producing an emergent industrial formation of entertainment technology that accommodates the imagined certainty of infrastructure and the entrepreneurial fuel of innovation.

The atmosphere of transmedia, platform plenitude, and branded viewing experiences did not naturally evolve from primitive primordial media to a beautifully cross-pollinated hybrid. (Let's just say the jury is out on whether it's a star-baby or a platypus.) There is certainly considerable stasis amid all this talk of upheaval, reiterated as it is over and over again. This atmosphere emerges from a long history of efforts to connect the entertainment business with the electronics industry and features the audiovisual infrastructure that an influential wired class lives with and imagines is part of all forms of contemporary human society. Our audiovisual infrastructure descends from concrete economic priorities that are inseparable from conceptual frames. Consumer electronics relies on a language of transferability and interchangeability. This language has a notable visual dimension; the iconography of the ubiquitous magical touch screen, in particular, figures ghostly gestures of control as the conjuring rituals for media transference. But this ideational dimension of media operations is not translated directly into practice. Our media environment emerges, expands, and circulates with the concerted intellectual labor to explain it, interpret it, act upon it, and forecast future prospects. The entertainment technology summits are visible locations for this labor. The pursuit is freighted with an ideological task: to make essential the inessential and to conceal the operations of that naturalizing process. This operation is consequential, for it shapes and furthers the platform and device plenitude that marks our time. From touch-screen tablets to data-cloud lockers, the material conditions of our audiovisual experience constitute a field upon which value and distinction emerge. Attention to these conditions begins a process of exposing the mechanisms of technological and cultural power. Without this attention, media scholars risk becoming accidental spin doctors for an audiovisual infrastructure.

Appendix

Table 11.1. Selective Chronology of Entertainment Technology Industry Events, 2009–2013

Event	Date m/d/y	Presentation and Panel Titles, selected	Keynotes or Prominent Speakers	Event Producers and Sponsors (featured)
3D Entertainment Summit	9/16–17/09	"Business side of 3D film"; "3D Alternate Content: Live Events, Music and Sports"	Jeffrey Katzenberg (CEO and Director, DreamWorks Animation SKG), Michael Lewis (Chair and CEO, RealD), Sandy Climan (CEO, 3ality Digital), Peter Bart (President and Editorial Director, *Variety*)	RealD, Sony, 3ality, IMAX, Panasonic, Virtual Images
Entertainment and Technology Summit	5/3/10	"Consumer's Perspective—Research Conducted at Sony/CBS 3D Research Center, Las Vegas"; "Technology: The Big Picture"	Kevin Beggs (President, Lionsgate Television), David Poltrack (Chief Research Officer, CBS Corporation, and President, CBS Vision), Daniel Scheinman (Senior Vice President and General Manager, Cisco Media Solutions Group)	Cisco Eos, nuMetra, *Variety*, Digital Hollywood
3D Entertainment Summit	9/15–16/10	"Home 3D—Solutions and Strategies"; "Turning the Corner from SD to HD to 3D"	Jeffrey Katzenberg (CEO and Director, DreamWorks Animation SKG), M. Night Shyamalan (Writer, Director), Chris Cookson (President, Sony Pictures Technologies/Sony Pictures Entertainment), John Rubey (President, AEG Network LIVE)	RealD, Sony, Sony Digital Cinema
Entertainment and Technology Summit Fall	10/18/10	"Who is King of the Digital Living Room?"; "State of the TV Business"; "Navigating Social Media for Film and TV"	Michael Eisner (former CEO, Walt Disney), Matt Diamond (CEO, Alloy Media Marketing), Hardie Tankersley (Vice President Innovation, Fox Broadcasting), Michael Kassan (CEO/Chair, MediaLink)	*Variety*, Digital Hollywood, Consumer Electronics Association (CEA), SAP, Cinram, Scenechronize, Alphabird, Tremor Media, Dolby, Cisco Eos, 3 Audio Alliance, AOL, thismoment

Event	Date m/d/y	Presentation and Panel Titles, selected	Keynotes or Prominent Speakers	Event Producers and Sponsors (featured)
Future of Film Summit	11/9/10	"Kickstarter SuperSession: Is Crowdsourcing the Future of Film Production and Marketing?"; "Why do Women Matter in Hollywood? Their Influence behind the Camera, in front of the Camera, and in the Audience"	Stacy Snider (Co-chair and CEO, DreamWorks Studios), Jon Landau (Producer, Lightstorm Entertainment), David Cohen, (Associate Editor, *Variety*)	Alphabird, BSAT Labs, CEA, Cinedigm, Cinram, Cisco Eos, Dolby, Rovi, Scenechronize, Visible Measures, *Variety*, Digital Wire Media
Variety Home Entertainment Hall of Fame Awards	12/6/10	n/a	Inductees: Ted Sarandos (Netflix), Bill Clark (Anchor Bay Entertainment), Marty Gordon (Philips Electronics), Andy Parsons (BDA Promotions) accepting for BDA	*Variety*, CEA, and *Home Media* Magazine
LA Mobile Entertainment Summit	12/7–8/10	"Marketing Entertainment Through Mobile: How Hollywood Uses Mobile as a Marketing Tool for Feature Films and TV"; "Consumer-Generated Media and User-Generated Content on Mobile: It's Big, It'll Get Bigger: Is it a Threat or a Boon to Professional Content Creation?"	Cameron Clayton (Senior Vice President Mobile and Digital Applications, The Weather Channel), Michael Scogin (Vice President Wireless MTV), John Kosner (Senior Vice President, ESPN Digital Media), Robert Gelick (Senior Vice President, CBS), Peter Guber (CEO, Mandalay Entertainment), David Hill (CEO, Fox Sports Television Group), Christoph Hoerenz (Chief Marketing Officer, Fox Mobile Group), Tim Kring (Producer and "Master Storyteller"), Brian Seth Hurst (CEO, The Opportunity Management Company), Natalie Farsi (Head of Mobile, Warner Bros.)	The Weather Network, ESPN, Unicomm

continued

Table 11.1. (continued)

Event	Date m/d/y	Presentation and Panel Titles, selected	Keynotes or Prominent Speakers	Event Producers and Sponsors (featured)
Entertainment Content Protection Summit	12/8/10	"Al Queda and McDonald's: What They Have in Common"	Taylor Hackford (Director/Producer and President, Directors Guild of America), Richard Atkinson (Chief Piracy Specialist and Consultant, Anti-Piracy Worldwide and Conference Chair), Joel Bigley (Senior Vice President, Worldwide Operations, Deluxe Digital Studios), Kaye Cooper-Mead (Executive Vice President, Worldwide Distribution, Summit Entertainment)	2011 International CES (Consumer Electronics Show), Civolution, Deluxe, Rovi, AFTRA, Irdeto, Peer Media Technologies, Fortium, Sony DADC
Entertainment Matters	1/6–9/11	n/a	David Kenny (President, Akamai), Joseph V. Tripodi (Executive Vice President, Coca-Cola Company), Michael Roth (CEO, InterPublic Group), Sir Martin Sorrell (CEO, WPP)	*Variety*, Cricket Wireless, CEA
Hollywood IT Summit	3/17/11	"Universal Media ID: The Rosetta Stone of Ubiquitous Trading"; "Digital Data Reporting: Chaos to Calm"; "Navigating the Outsourcing Maze"	Jeff Mirich (Walt Disney Studios and Conference Chair), Devendra Mishra (MESA and Conference Chair), Steve Andujar (Sony Pictures Entertainment), David Cohen (*Variety*), Linda Livingstone (Pepperdine University), Steve Weinstein (MovieLabs), Peter DeLisi (Santa Clara University)	SAVVIS, Hollywood IT Society, Media and Entertainment Services Alliance (MESA), *Variety*
Entertainment and Technology Summit	5/2/11	"State of the Entertainment Content Business"; "The Big Web Revamp: Where Does Hollywood Fit In?"; "A Conversation With Sumir Meghani, Groupon"	Sean Bailey (President, Walt Disney Studios Motion Picture Productions)	*Variety*, Digital Hollywood, thismoment, MediaLink, Greenberg Glusker, AFTRA, Microsoft, Facebook

Event	Date m/d/y	Presentation and Panel Titles, selected	Keynotes or Prominent Speakers	Event Producers and Sponsors (featured)
LA Mobile Entertainment Summit	9/20–21/11	"New Storytelling Machines: Entertainment in the Age of Mobility"; "Marketing Entertainment in a Mobile World"; "M-Commerce: Monetizing Mobile Entertainment"; "Tablets on the Rise"; "Gamification: How Mobile is Changing the Way Consumers Game"; "Mixing Fun and Profit: Mobile Ads and Entertainment"	Gene Simmons ("Rock God, Media Mogul, Multi-Hyphenate and Orstbo.com Spokesperson"), Barry Cottle (EVP/GM, Electronic Arts Interactive), David Lucatch (President, Orstbo.com), Alex Barkaloff (Executive Producer, Digital Media, Lionsgate)	Ortstbo, Blue Calypso, Moco Space, SRS Lab, Epicmobile, thismoment
3D Entertainment Summit	9/20–22/11	"SimulView: New Technology That Allows 3D on Playstation"	James Cameron and Vince Pace (President and CEO, Cameron/Pace Group), Jim Chabin (President and CEO, International 3D Society), Tom Rothman (President and CEO, Fox Filmed Entertainment), Bryan Burns (Vice President, ESPN), Michael Lewis (Chair and CEO, RealD), John Revie (Senior Vice President, Home Entertainment Product Marketing, Samsung), Chris McGurk (Chair and CEO, Cinedigm Digital Cinema) Mick Hocking (Vice President, Sony Computer Entertainment), Bob Dowling (Conference Chair)	RealD, LG Cinema 3D, Samsung
Film Marketing Summit	10/4–5/11	"State of the Business"	Ivan Reitman (Director/Producer), Tom Sherak (President, Academy of Motion Picture Arts and Sciences)	Variety, Stradella Road, Creative Impact Agency, CEA, Fizziology, Hilton Hotels, thismoment, Tremor Video, Yahoo, Film Funds, Connectivision, DG Entertainment

continued

Table 11.1. (continued)

Event	Date m/d/y	Presentation and Panel Titles, selected	Keynotes or Prominent Speakers	Event Producers and Sponsors (featured)
Digital 25: Leaders in Emerging Entertainment	10/17/11	n/a	Meyer Shwarzstein (Brainstorm Media), Jon Kirchner (DTS), Susan Margolin and Steve Savage (New Video), John Calkins (Sony Pictures Home Entertainment), Ken Ralston (Sony Pictures Imageworks), Thierry Coup, Chip Largman, Dale Mason and Mark Woodbury (Universal Creative)	Producers Guild of America
Entertainment and Technology Summit	10/17/11	"The State of the TV Business"; "Original Web Content Grows Up"; "Trendsetters in Entertainment Distribution"	Jon Favreau (Director/Actor), Kevin Mayer (Disney)	*Variety*, Digital Hollywood
Film Technology Summit	11/7/11	"State of Technology in Film: From Production to Exhibition"; "How Are the Roles of the Film Crew Impacted in the Digital Age?"; "What's Next for CGI Animation?"; "Understanding the New Technology Landscape: 4K, 3D, Higher Frames and More"; "Film Restoration Supersession"; "Digital Cinema Exhibitors: Technology and New Revenue Streams"; "Making Movie Stars out of Special Effects and Unique Imagery"	Shawn Levy (Producer/Director), Roland Emmerich (Producer/Director/Writer)	Sony, Digital Media Wire, Alphabird, Cinram, Cisco Eos, Dolby
Future of Film of Summit	11/8/11	"State of the Film Industry"; "What is the Outlook for Film Finance?"; "The Digital Explosion: Navigating the Best of Alternative Film Delivery and Marketing Models"	Ryan Kavanaugh (CEO, Relativity Media)	CEA, Fizziology, Cinedigm

Event	Date m/d/y	Presentation and Panel Titles, selected	Keynotes or Prominent Speakers	Event Producers and Sponsors (featured)
Entertainment Apps Conference	12/1/11	"I Want My App TV"; "New TV Viewing Party"; "Movie Apps and Social Networks"; "Monetizing the Second Screen"; "Getting Noticed: Apps and Content Discovery"	Scott Maddux (Senior Vice President, Product Development, Nielsen), Colin Helms (Vice President, MTV Digital Media, Viacom Media Networks), Ajay Shah (CEO, TV-plus)	BluFocus, Slacker, CEA, Mobovivo, 1K Studios, AFTRA
Variety Home Entertainment Hall of Fame Awards	12/5/11	n/a	Inductees: Brian Dunn (Best Buy), John Marmaduke (Hastings Entertainment), Gary Shapiro (CEA), Nintendo of America, Harry Potter (Film Franchise)	*Variety*
Content Protection Summit 2011	12/8/11	"Case Study: DVD 2/3 Investigation"; "Tech Deep-Dive: Latest Developments in Technological Tools for Battling Content Piracy"	Nicholas Chartier (Producer), Avi Lerner (Co-Chair, Nu Image)	Deluxe, Irdeto, Akamai, Civolution, *Variety*, CDSA, MESA
Entertainment Matters	1/10–13/12	"Merging Content with New Technologies: Content Reinvention with the Next Generation of Tablets, eReaders and Mobile Devices"; "Multi-Screen Universe"; "Video Anytime Anywhere: Video Across Platforms"; "The State of Games Industry"; "What's Cooking? Technologists Eye the Future of Entertainment"	Ambassador: Eliza Dushku (Actress) Keynote: Robert Kyncl (YouTube)	*Variety*, United Talent Agency, Ericsson, CEA

continued

Table 11.1. (continued)

Event	Date m/d/y	Presentation and Panel Titles, selected	Keynotes or Prominent Speakers	Event Producers and Sponsors (featured)
2nd Screen Summit	2/22/12	"TV Goes Social"; "Is 2nd Screen Mass Market?"; "Creating an Engaging Experience for 2nd Screen Consumer Applications"; "Metadata: Evolving 2nd Screen Experiences"; "Content Licensing: Mastering The Cross-Platform, Mutli-Screen Mix"	Bill Baxter (CTO, BuddyTV)	MESA, CEA, Technicolor, 1K Studios
Hollywood IT Summit	3/2/12	"The New Rules of Building an Enterprise Software Company That People Love"	Devendra Mishra (Chief Strategist, MESA and Conference Chair), Steve Andujar (Vice President and CIO, Sony Pictures), John Herbert (EVP and CIO, Twentieth Century Fox), Steve Lapinski (SVP, IT, NBC Universal)	Pricewaterhouse Coopers (PwC), Teradata QlikView, Mark Logic, Oracle, MESA
The TV Summit	3/20/12	"State of the Film Industry"; "Facts on Pacts"; "Showrunners SuperSession"	Dana Walden (Chair, Twentieth Century Fox TV), Gary Newman (Chair, Twentieth Century Fox TV, Bill Abbott (President and CEO Crown Media Family Networks [Hallmark Channel/Hallmark Movie Channel]); David Eilenberg (Head, Development and Current Programming, Mark Burnett Productions)	FreeWheel, Ipsos MediaCT, Civolution, AFTRA, Yahoo!, Milyoni, Mobovivo, Beaulieu Vineyard, Tremor Video, BrightRoll, Cartoon Network

Event	Date m/d/y	Presentation and Panel Titles, selected	Keynotes or Prominent Speakers	Event Producers and Sponsors (featured)
Entertainment and Technology Summit	4/30/12	"State of the Entertainment Content Business"; "The Arrival of Advertising and Multi-Platform Online Video"; "The YouTube Effect: Hyper-Channelization and the New Programming Landscape"; "It's a Jungle Out There: Making Sense of Mobile, Smart TV, Over-the-Top, Over-the-Air"; "New Entertainment Content: Apps, Social Gaming, and Commerce"	Damon Lindelof (Writer/Producer), Gale Anne Hurd (Executive Producer), Michael Kassan (Chair and CEO, MediaLink)	Digital Hollywood, Yahoo!, BrightRoll, FreeWheel, PWC, Newvue, SAG—AFTRA One Union, Skype, Gracenote, Northhighland (Worldwide consulting), Akamai, 1K Studios (Cinram Digital Media), GroundLink
Massive: The Advertising Summit	6/7/12	"Advertiser Trendsetters: How Brands Are Breaking Through and Succeeding in Media"; "The New Multi-Platform Opportunity: Moving Advertising Across Mobile, VOD, Online Video and More"	Anne Sweeney (Co-Chair, Disney Media Networks and President, Disney/ABC Television Group), Frank Cooper (CMO, Global Consumer Engagement, Pepsi), Nigel Morris (President and CEO, Aegis Media North America), Amy Powell (President, Insurge Pictures and Paramount Digital Entertainment)	Social Vibe, GroundLink, Live Nation, SAP, Google, The Jim Henson Company, House, Co.Create, Fast Company, SID, Mobovivo, Warner Cable Media, Visible Measures, The One Club, Phenomblue, Latino
Venture Capital New Media Summit	6/27/12	"Entertainment Game-Changers: Who Is Transforming Content and Consumer Experience for the Better?"; "The Future of Social Media: Gaming, E-Commerce, Life-sharing—What Are Ways Still to Innovate?"; "Cult of Celebrity: When Stars Align with Technology Companies"	Ben Silverman (CEO, Electus), Antonio Villaraigosa (Mayor, Los Angeles), Todd Lutwak (Partner, Andreessen Horowitz), Todd Simon (Head of Digital Media Investment Banking, Oppenheimer)	Shamrock Capital Advisors, KPMG, SNL Kagan, Cinedigm, Watchwith, Hark, The Churchill Club, Boogar Lists, Founders Space, Plug and Play, Keiretsu Forum

continued

Table 11.1. (continued)

Event	Date m/d/y	Presentation and Panel Titles, selected	Keynotes or Prominent Speakers	Event Producers and Sponsors (featured)
MultiScreen Summit	9/19–20/12	"Marketing and Advertising Content Across Platforms"	Kevin Morrow (Vice President Strategic Partnerships, Samsung Media Center Solutions), Coleman Breland (COO, Turner Broadcasting), Tim Kring (Master Storyteller), Brian Seth Hurst (CEO, Opportunity Management Company), Chris McGurk (CEO, Cinedigm), Jordan Levin (President, Alloy Digital)	Samsung, Ortsbo
3D Entertainment Summit	9/19–20/12	"New Global 3D Market Place"; "3D Hollywood Masters: What Makes Good 3D?"	Recipient of the inaugural 3D Visionary Award, Jeffrey Katzenberg, CEO, DreamWorks Animation SKG via "an exclusive one-on-one interview shot in 3D"	RealD, Ultra D
Film Technology Summit	9/20/12	"The New Technology Playground: I-Filmmaking, Shooting in 4K—What's Next on the Set?"; "State of Film Production in the Digital Age"; "The Arrival of Higher Frame Rates: Are We Ready?"; "Beyond the Bang: Masters of Creativity in Special Effects"	Joe Letteri (Partner, Weta Digital and Visual Effects Supervisor), Ty Warren (Executive Vice President, Physical Production, Legendary Pictures), Darcy Antonellis (President, Technical Operations, Warner Bros.), Victoria Alonso (Executive Vice President, Visual Effects and Post-Production, Marvel Studios)	Sony Digital Cinema, MIDI Screens, Aspera, Ortsbo, RealD, Samsung, Ultra-D
HITS Digital Marketing and Analytics	9/28/12	"METADATA: Trigger for Monetization!"; "Media Measurement: In Search of the Holy Grail"; "HITS-DMA Summit: Technology Spotlights"	Dwight Caines (President, Worldwide Digital Marketing, Sony Pictures Entertainment)	Aspera, PwC, NetBase/SAP, Cloudera, MapR, Show of Hands, Cinedigm
Variety Film Marketing Summit	10/24/12	"Behind Every Blockbuster Is a Good Social Marketer"	Eli Roth (Writer/Producer/Director)	Tremor Video, Yahoo, SAG-AFTRA, AOL, CEA, Simulmedia, TouchTunes, *Variety*

Event	Date m/d/y	Presentation and Panel Titles, selected	Keynotes or Prominent Speakers	Event Producers and Sponsors (featured)
Entertainment Apps Conference	11/29/12	"Smart TVs and Connected Devices: How Many Platforms are Too Many?"	Tom Engdahl (President and CEO, Magic Ruby), Woody Sears (Founder, Zuuka), Jim Molinets (Vice President and General Manager, Mobile Games, Glendale Disney Interactive Studios), Henry Oh (Director, Strategic Initiatives, Animoca)	The Echo Nest, Tapjoy, Newvue, 1K Studios, SAG-AFTRA
Variety Home Entertainment Hall of Fame Awards	12/3/12	n/a	Inductees: Avi Lerner (Nu Image and Millennium Films), Mitch Singer (Sony Pictures Entertainment and Digital Entertainment Content Ecosystem), Bill Sondheim (Gaiam), Star Wars: The Complete Saga	Yahoo, CEA, Home Media, MESA, DEG, SAG-AFTRA, *Variety*
Future of Film Summit	12/5/12	"Facts on Pacts Super Session: What is the Future of Studio Production Deals?"	Kevin Smith (Director-Writer), Ralph Garman (KROQ Radio Personality and Actor), Danielle De Palma (Senior Vice President, Digital Marketing, Lionsgate), Alex Barkaloff (US/EMEA Head of Tom Hanks's "Electric City")	BitTorrent, SAG-AFTRA, Oakwood Worldwide, International CES
Content Protection Summit	12/6/12	"Technology Strategies for Protecting Content and Growing the Home Entertainment Business"	Christopher J. Dodd (Chair and CEO, Motion Picture Association of America), General (Ret.) James Cartwright (Harold Brown Chair in Defense Policy Studies, Center for Strategic and International Studies)	Deluxe, Akamia, Irdeto, SAG-AFTRA
CES Entertainment Matters	1/9–10/13	"Digital Content Big Leagues"; "The Future of Television"	Anthony Bay (Vice President, Worldwide Video, Amazon.com), Morgan Spurlock (Writer/Director/Producer), Hanno Basse (CTO, 20th Century Fox)	PwC, ICM Partners, FreeWheel, Commerce Hotel Casino, *Variety*, CEA

Notes

1. Chris Anderson, *The Long Tail: Why the Future of Business Is Selling Less of More* (New York: Hyperion, 2006). Comments from Caetlin Benson-Allott, Lisa Parks, and Nicole Starosielski greatly assisted the development of this chapter. Invaluable research assistance came from Julio Valdez and Ashley McAskill.

2. Thomas Schatz, "The Studio System and Conglomerate Hollywood," in *The Contemporary Hollywood Industry*, ed. Paul MacDonald and Janet Wasko (Oxford: Blackwell, 2008), 11–42.

3. For more on the concept of the "technological tentpole," see Charles R. Acland, "The End of James Cameron's Quiet Years," in *Media Studies Futures*, ed. Kelly Gates, *The International Encyclopedia of Media Studies Volume VI*, general ed. Angharad N. Valdivia (London: Blackwell, 2013), 269–95.

4. "Industry Leaders Forecast Dramatic 3D Expansion at Third Annual 3D Entertainment Summit," *Business Wire*, September 22, 2010.

5. Several of these terms emerge from the influential work of Henry Jenkins, including *Convergence Culture: Where Old and New Media Collide* (New York: New York University Press, 2006) and, with Sam Ford and Joshua Green, *Spreadable Media: Creating Value and Meaning in a Networked Culture* (New York: NYU Press, 2013).

6. For more on this, see Charles R. Acland, "IMAX Technology and the Tourist Gaze," *Cultural Studies* 12, no. 3 (1998): 429–45. Paul Grainge studies this aspect of the movie business in his *Brand Hollywood: Selling Entertainment in a Global Media Age* (New York: Routledge, 2008).

7. Vicki Mayer, *Below the Line: Producers and Production Studies in the New Television Economy* (Durham, N.C.: Duke University Press, 2011), 31–65.

8. Richard Maxwell and Toby Miller, *Greening the Media* (New York: Oxford University Press, 2012).

9. Lisa Parks, " 'Stuff You Can Kick': Toward a Theory of Media Infrastructures," in *Between Humanities and the Digital*, ed. David Theo Goldberg and Patrik Svensson (Cambridge, Mass.: MIT Press, 2015), 355–73.

10. Consumer electronics has evolved to embrace information/communication technology such that it may no longer make sense to consider them distinct industrial domains. For this reason, Maxwell and Miller (*Greening the Media*) prefer the term CE/ICT.

11. Tarleton Gillespie, "The Politics of 'Platforms,' " *New Media and Society* 12, no. 3 (2010): 352.

12. Ibid., 359.

13. José van Dijck, *The Culture of Connectivity: A Critical History of Social Media* (New York: Oxford University Press, 2013), 4.

14. Ibid., 27 (original emphasis).

15. Ibid., 6.

16. See for example Jonathan Sterne, *MP3: The Meaning for a Format* (Durham, N.C.: Duke University Press, 2012); Nicole Starosielski, " 'Warning: Do Not Dig': Negotiating

the Visibility of Critical Infrastructures," *Journal of Visual Culture* 11, no. 1 (2012): 38–57; and Lisa Parks, "Around the Antenna Tree: The Politics of Infrastructural Visibility," *FlowTV*, 9, no. 8 (2009).

17. Geoffrey C. Bowker, Karen Baker, Florence Millerand, and David Ribes, "Toward Information Infrastructure Studies: Ways of Knowing in a Networked Environment," in *International Handbook of Internet Research*, ed. Jeremy Hunsinger, Lisbeth Klatstrup, and Matthew Allen (New York: Springer, 2010), 98 (original emphasis).

18. Susan Leigh Star, "The Ethnography of Infrastructure," *American Behavioral Scientist* 43, no. 3 (1999): 381–82.

19. Lisa Gitelman, *Always Already New* (Cambridge, Mass.: MIT Press, 2007).

20. Gillespie, "The Politics of 'Platforms,'" 359.

21. Marc Graser, "Hollywood Wired for Business: CES Showcases Latest Hardware for Content Delivery," *Variety*, January 16–22, 2012, 7.

22. "Digital Hollywood at CES to Explore New Digital Content and Market Opportunities," *PR Newswire* (New York), October 17, 2000.

23. J. D. Lasica, *Darknet: Hollywood's War against the Digital Generation* (Hoboken, N.J.: Wiley, 2005).

24. Sea-Jin Chang, *Sony vs. Samsung: The Inside Story of the Electronics Giants' Battle for Global Supremacy* (Toronto: Wiley, 2008), 12.

25. Toshio Mitsufuji, "Strategic Alliance in the Video Industry: The Acquisition of Columbia by Sony and MCA by Matsushita." *Journal of Strategic Change* 2 (1993): 207–14.

26. Lasica, *Darknet*, 23.

27. See especially John Caldwell, *Production Culture: Industrial Reflexivity and Critical Practice in Film and Television* (Durham, N.C.: Duke University Press, 2008).

28. Ibid., 104.

29. Ibid.

30. "Variety and Digital Hollywood Launch the Entertainment and Technology Summit," *Business Wire* (New York), April 22, 2010.

31. "Variety Announces the Entertainment and Technology Summit Lineup," *Business Wire*, March 17, 2011.

32. Ibid.

33. *Digital Hollywood*, ed. Victor Harwood, http://www.digitalhollywood.com.

34. "2012 MultiScreen Summit Announces Top Speakers as Event Explores New Screen Economy," *Business Wire*, August 30, 2012.

35. See http://www.digitalhollywood.com.

36. This is most fully elaborated in Pierre Bourdieu, *Distinction: A Social Critique of the Judgement of Taste* (1979; Cambridge, Mass.: Harvard University Press, 1984) and Pierre Bourdieu, *The Field of Cultural Production* (1986; New York: Columbia University Press, 1993).

37. "Producers Guild of America to Hold Annual 'Produced by Conference' June 8–10 2012," *Consumer Electronics*, Arlington, Virginia, December 19, 2011.

38. "Digital Hollywood at CES."

39. Ibid.

40. Ibid.

41. Chang, *Sony vs. Samsung*, 13.

42. "2006 International CES Rocks Las Vegas with Hot New Technology, Cool Products and Famous Faces; Record-Breaking Show Opened with 1.6 Million Net Square Feet of Exhibit Space, 2,500 Exhibitors and Industry Leading Speakers," *Business Wire*, January 7, 2006.

43. Paul Boutin, "Live Coverage of Sony's Sir Howard Stringer," *engadget.com*, January 5, 2006.

44. Josef Adalian, "Sony Sparks to CES Confab," *Daily Variety*, November 14, 2007, 29.

45. Graser, "Hollywood Wired for Business," 7.

46. Ibid.

47. "Entertainment Matters Brings Hollywood to the 2011 CES," *Consumer Electronics*, Arlington, Virginia, October 21, 2010.

48. Graser, "Hollywood Wired for Business," 7.

49. "Eliza Dushku Named Abassador for 2012 CES Entertainment Matters Program," *Consumer Electronics*, October 27, 2011.

50. "*Variety* Announces Sponsorship of the Entertainment Matters Program, Entertainment Matters Daily and Two Entertainment Summits All at the 2013 International CES," *Business Wire*, September 11, 2012.

51. Graser, "Hollywood Wired for Business," 7.

52. "Musical Icons, Hollywood Stars and TV Personalities to Attend 2012 International CES," *Business Wire*, January 3, 2013.

53. "*Variety* Announces Sponsorship."

54. "Entertainment Matters: Getting to Know You," *Consumer Electronics*, January 6, 2013.

55. Ibid.

56. Ibid.

57. "CES 2013: Keynote Conversation: Cross-Platform Storytelling with Tim Kring," *TransMedia.ca*, January 4, 2013.

58. "Entertainment Matters at the 2013 International CES," *Consumer Electronics*, January 6, 2013.

59. "Trio Tapped for Variety Home Entertainment Hall of Fame," *Variety.com*, October 11, 2010.

60. Advertisement for "Entertainment Matters 2011," *Variety*, January 3–9, 2011, 32 (back cover).

Contributors

CHARLES R. ACLAND is professor and Concordia University research chair in Communication Studies in Montreal. His books include *Screen Traffic: Movies, Multiplexes, and Global Culture* (Duke University Press, 2003), *Swift Viewing: The Popular Life of Subliminal Influence* (Duke University Press, 2012), and *Useful Cinema* (Duke University Press, 2011), co-edited with Haidee Wasson. He is co-director of the Media History Research Centre at Concordia University.

PAUL DOURISH is professor of Informatics in the Donald Bren School of Information and Computer Sciences at UC Irvine, with courtesy appointments in the Departments of Computer Science and Anthropology; he co-directs the Intel Science and Technology Center for Social Computing. His research focuses on information technology as a site of social and cultural production, combining topics in human-computer interaction, ubiquitous computing, and science and technology studies. Dourish is the author of *Where the Action Is: The Foundations of Embodied Interaction* (MIT Press, 2001) and, with Genevieve Bell, *Divining a Digital Future: Mess and Mythology in Ubiquitous Computing* (MIT Press, 2011).

SARAH HARRIS earned her PhD in the Department of Film and Media Studies at UC Santa Barbara in 2013. She has research interests in global ICT policy, cultures of circumvention, and Internet access and activism. Harris's dissertation, "Proxy Cultures: Circumvention in Turkish Information Society," was based on more than two years of ethnographic fieldwork in Turkey and was supported by fellowships from the Institute of International Education, American Research Institute of Turkey, Institute of Turkish Studies, and the University of California. Harris is currently pursuing an MBA at the Kellogg School of Management at Northwestern University, with a focus on high-tech industries and their social impacts.

JENNIFER HOLT is associate professor of Film and Media Studies at the University of California, Santa Barbara. She is the co-editor (with Alisa Perren) of *Media Industries: History, Theory, and Method* (Blackwell, 2009) and (with Kevin Sanson) *Connected Viewing* (Routledge, 2013), and author of *Empires of Entertainment* (Rutgers University Press, 2011), which examines deregulation and media industry conglomeration from 1980–1996. She is also director of the Carsey-Wolf Center's Media Industries Project and a founding member of *Media Industries* journal's editorial collective.

SHANNON MATTERN is associate professor in the School of Media Studies at The New School. She has written about archives, libraries, and other media-architectures, media infrastructures, place branding, public design projects, multisensoriality, and media exhibition. She is the author of *The New Downtown Library: Designing with Communities* (University of Minnesota Press, 2007) and numerous articles and book chapters. Her research has been supported by the Mellon Foundation, the Graham Foundation for Advanced Studies in the Fine Arts, the Canadian Centre for Architecture, and the Korea Foundation. You can find her at wordsinspace.net.

TOBY MILLER is the Sir Walter Murdoch professor of Cultural Policy Studies at Murdoch University and professor of Journalism, Media and Cultural Studies at Cardiff University/Prifysgol Caerdydd. The author and editor of more than thirty books, Miller's work has been translated into Spanish, Chinese, Portuguese, Japanese, Turkish, German, and Swedish. His two most recent volumes are *Greening the Media* (with Richard Maxwell) and *Blow Up the Humanities* (both 2012). He can be contacted at tobym69@icloud.com and his adventures scrutinized at www.tobymiller.org.

LISA PARKS is professor of Film and Media Studies and director of the Center for Information Technology and Society at the University of California, Santa Barbara. Parks is the author of *Cultures in Orbit: Satellites and the Televisual* (Duke University Press, 2005), and *Media Spaces and Global Security: Coverage since 9/11*

(Routledge, forthcoming) and is working on another book titled *Mixed Signals: Media Infrastructures and Cultural Geographies*. She is also co-editor of *Down to Earth: Satellite Technologies, Industries and Cultures* (Rutgers University Press, 2011), *Undead TV* (Duke University Press, 2007), and *Planet TV* (New York University, 2003).

CHRISTIAN SANDVIG is the Steelcase Research Professor and associate professor in the Department of Communication and the School of Information at the University of Michigan. His research investigates the societal implications of communication and information technology infrastructure. He is the author of numerous peer-reviewed articles and book chapters.

NICOLE STAROSIELSKI is assistant professor in the Department of Media, Culture, and Communication at New York University. Her research focuses on the global distribution of digital media, and the relationships between technology, society, and the aquatic environment. In *The Undersea Network* (Duke University Press, 2015), Starosielski charts the historical, cultural, and environmental dimensions of transoceanic cable systems.

JONATHAN STERNE is professor and James McGill chair in Culture and Technology in the Department of Art History and Communication Studies and the History and Philosophy of Science Program at McGill University. He is author of *MP3: The Meaning of a Format* (Duke University Press, 2012), *The Audible Past: Cultural Origins of Sound Reproduction* (Duke University Press, 2003), and numerous articles on media, technologies, and the politics of culture. He is also editor of *The Sound Studies Reader* (Routledge, 2012). His new projects consider instruments and instrumentalities; histories of signal processing; and the intersections of disability, technology, and perception. Visit his website at http://sterneworks.org.

HELGA TAWIL-SOURI is associate professor in the Department of Media, Culture, and Communication at New York University, where she is also director of the Kevorkian Center for Near Eastern Studies. Her scholarship focuses on spatiality, technology, and politics in the Middle East, with a focus on the role of technology and culture in everyday experiences in Palestine-Israel. Her work looks at the intersections of media spaces such as the Internet, telecommunications, and film, and such physical boundaries as checkpoints, barriers, and ID cards. Helga also writes about new technologies and their relationship to political and economic transformations in Arab media. She is also a photographer and documentary filmmaker.

PATRICK VONDERAU is associate professor and senior lecturer at the Department of Media Studies at Stockholm University. His most recent book publications include *Behind the Screen: Inside European Production Cultures* (Palgrave, 2013, with Petr Szczepanik); *Moving Data: The iPhone and the Future of Media* (Columbia University Press, 2012) and *The YouTube Reader* (Wallflower, 2009) (both with Pelle Snickars); and *Films That Work: Industrial Film and the Productivity of Media* (Amsterdam University Press, 2009, with Vinzenz Hediger). Patrick is a co-editor of Germany's leading media studies journal, *Montage*, and a cofounder of NECS— European Network for Cinema and Media Studies (www.necs.org).

Index

energy use: data centers and, 1–2, 82–84, 92n58, 143; economic considerations, 66; increases in, 79, 121, 141; rural Zambia Internet infrastructure and, 120–22. *See also* environmental impact

Ensmenger, Nathan, 227

entertainment, study of, 5

Entertainment and Technology Summit, 256, 266, 268, 270, 273

Entertainment Apps Summit, 255, 271, 275

entertainment industry, 252–75; background/history, 252–54; chart of industry summits, 266–75; cross-promotion, 21, 246–47, 253; film history, 97; film production and cable systems, 58–59, 59–60, 69n10; industry events in general and, 254–57, 263–65; intermedia relations and, 21, 248–49, 256, 259–60, 261–62, 263. *See also* video distribution

Entertainment Matters, 260–61, 261–62, 263, 268, 271, 275

environmental impact, 14, 137–50; e-waste, 116–17, 139–41, 142–43; e-waste art and, 146–50; Google's marketing and, 73–74, 75–76; obsolescence, 146–47; production of electronics, 138, 250; undersea cables, 4. *See also* energy use; natural resources/systems

epigraphy, 100–101, 103, 104, 106, 107

Ernst, Wolfgang, 85

e-waste art, 19, 137, 146–50

Exterior Gateway Protocol (EGP), 20, 194–95, 199, 200, 204n20

Facebook, 71, 72, 84, 139

Fanon, Frantz, 34

Fast Company (magazine), 140

Fatpipe, the, 58, 59, 65

Favro, Diane, 99

Fiji, cable system development in, 58, 59

film, urban history and, 97

Forbes (magazine), 140

Foreign Policy (magazine), 146–47

Fowler, Mark, 79

freedom of speech, 74, 209. *See also* censorship/surveillance; decentralization

Friedberg, Anne, 32

Friedman, Yona, 146

Frost, Samantha, 5, 10

Fuller, Matthew, 82

Future Film Summit, 255, 267, 270, 275

Galloway, Alexander, 8, 184, 185, 192

gaming, 61, 67, 239, 250, 256

gateways (routers), 67; centralization and, 199; defined, 190; prices, 227; routing tables and, 190–91, 192, 193, 194, 195–97; transmission speed and, 62. *See also* routing

Gaza Strip, the, 159, 160, 162, 164–65, 179n18

Geertz, Clifford, 94

gender discourse: early Internet design and, 227–28; entertainment industry events and, 263; rural Zambia, 123–24, 130–33; Turkey's cybercafés and, 205, 206, 217–20, 221

Gigabit Networking (Partridge), 188

Gillespie, Tarleton, 10, 250–51, 252

Gitelman, Lisa, 252

Goffey, Andrew, 82

Google: data center expansion, 82; data center marketing, 71, 72–74, 75–77, 80–81; energy use, 83, 84, 143; Hamina data center description, 1–2; transpacific cable, 60

Gordon, Eric, 97

Graham, Stephen: disruption, 13; maintenance and repair, 213; relationality and layering, 11, 103, 104; scale, 107; splintering of experience, 184

Graser, Marc, 253

Greening the Media (Maxwell and Miller), 14, 121, 250

Greenpeace, 82

Gregg, Melissa, 15

Guardian, The (newspaper), 63, 139

halftone printing, 46

Hall, Peter, 94

Harris, Sarah, 12, 20, 205, 279–80

Harrist, Robert, 101

Harvey, David, 48

THE GEOPOLITICS OF INFORMATION

Digital Depression: Information Technology and Economic Crisis   *Dan Schiller*
Signal Traffic: Critical Studies of Media Infrastructures   *Edited by Lisa Parks*
    and Nicole Starosielski

The University of Illinois Press
is a founding member of the
Association of American University Presses.

Composed in 10.5/13 Marat Pro
by Kirsten Dennison
at the University of Illinois Press
Manufactured by Cushing-Malloy, Inc.

University of Illinois Press
1325 South Oak Street
Champaign, IL 61820-6903
www.press.uillinois.edu